# THE DETAILS OF A CELL STRUCTURE

# THE DETAILS OF A CELL STRUCTURE

*Edited by*

**Dr. Kiran Abasaheb More**

Delve
PUBLISHING

www.delvepublishing.com

**The Details of a Cell Structure**

*Dr. Kiran Abasaheb More*

**Delve Publishing**
224 Shoreacres Road
Burlington, ON L7L 2H2
Canada
www.delvepublishing.com

Email: orders@arclereducation.com



ISBN: 978-1-77469-864-8 (Hardcover)

# ABOUT THE EDITOR

**Dr. Kiran Abasaheb More**, born in Bhogaon (Devi) Tahsil Jintur of Parbhani district of Maharashtra, India, in 1978, completed M.Sc in Botany, at Shree Shivaji College Parbhani, Maharashtra, Affiliated with Swami Ramanand Teerth Marathwada University Nanded, Maharashtra. The author also completed his Ph.D in 2010 in field of Taxonomy of Angiosperms and he joined the department of Botany, Yashwantrao Chavan Arts and Science Mahavidyalaya, Mangrulpir as an Assistant Professor. He is the Chief Editor of Research Journal Multilogic in science (Journal for Applied science), He edited more than 35 issues of research journals and authored 25 research papers. Dr. More is awarded by Young scientist award of Science and tech society for Integrated ruler improvement Telangana , and also awarded by Excellence in Teaching GENESIS ARBAN &RULER Devolopment Society at the occasion of ICAAASTSD International conference 2018. He is the editor of proceeding of International conference on Advances in Agriculture and Allied science technologies for sustainable Development Society Special issue Jan 2018 and he is a Life Member of "GUARD Society". He also chaired a number of National and International Conferences, he is also a Member of Botany teacher association of Sant Gadge baba Amravati University Amravati, Maharashtra, India.

# TABLE OF CONTENTS

# LIST OF FIGURES

# LIST OF TABLES

# PREFACE

With a total age of 3.8 billion years and counting, there is still much to learn about the principles of how life works. This approachable book takes you from the primordial soup to the present, showing how specialized cells gave rise to everything that is alive, from the tiniest amoeba to walking, talking humans. Whether you're enrolled in a cell or molecular biology course and require a concise overview, or you're simply curious about the newest discoveries, this completely revised edition is your all-access pass to our inner world.

Robert Hooke discovered the cell in 1665, and his book Micrographia describes it. He detailed 60 'observations' of diverse things using a crude compound microscope in this book. Very tiny shards of bottle cork provided one observation. Hooke discovered a slew of small holes dubbed "cells" by him. This is derived from the Latin words Cella, which meant 'a little apartment' where monks lived, and Cellulae, which meant a honeycomb's six-sided cell. Hooke, on the other hand, had no idea what they were made of or what they did. The empty cell walls of plant tissues were mistaken for cells by Hooke. Hooke was unable to discern that the cells he was seeing contained other interior components due to the poor magnification of microscopes available at the time. He didn't believe the "cellulae" were living because of this. The nucleus and other components found in most living cells were absent from his cell observations. Hooke also noticed mold on leather in Micrographia, which was a bluish tone. He couldn't find "seeds" that would have shown how the mold was multiplying in abundance after examining it under his microscope. As a result, Hooke hypothesized that spontaneous generation, whether caused by natural or artificial heat, was to blame. Others did not reject this ancient Aristotelian hypothesis because it was still acceptable at the time, and it was not discredited until Leeuwenhoek subsequently discovered that generation could be accomplished in a different way.

*The Details of a Cell Structure* decodes the language and theories that can be exhausting for even the most diligent student. It covers everything from fundamental ideas to how modern technology, genetic testing, and microarray techniques are expanding research and job opportunities. Additionally, it provides important ideas for preparing for – and acing — your examinations!

This book helps the readers:

Investigate the structure and function of cells—and discover why cellular context is critical for illness research.

Investigate how molecular biology can be used to address global issues.

Recognize how DNA determines features and how cells regulate it

Improve the knowledge and performance by utilizing online resources and study techniques.

This book contains information ranging from microscopic minutiae to macro themes and is equally beneficial for students, teachers, and scientists in the area of biology and medicine.

–Author

# CHAPTER 1

# FUNDAMENTALS OF CELL

CONTENTS

## 1.1 INTRODUCTION

A cell is a structure that consists of a volume of cytoplasm enclosed by a semi-permeable layer/wall known as the plasma membrane. A cell is the tiniest component capable of carrying out life's functions. It contains the cytoplasm, several cellular organelles, and the nucleus or nuclear material. Cells are the fundamental part of all living organisms, as well as all creatures are composed of at least one cell (Oshima & Campisi, 1991). Apart from sharing a similar fundamental structure, all cells undergo comparable life cycles. These involve material movement, energy acquisition, and use, sewage disposal, duplication, and environmental response. Cells are the simplest fundamental element of all living creatures. They carry out their operations autonomously; they reproduce and split. They are often referred to as life's basic components. Cell biology or cytology is the science that involves the analysis of cells. A human individual is believed to contain more than ten trillion cells (Sharaf & Orhan, (2014). It is 1013 cells numerically, and when seen via a microscope, they are invisible to the human eye. Cells are classed into two groups based on the organization of their membranes, the diversity and structure of their cytoplasmic organelles, and the complication of their nuclear area. These are called Prokaryotic and Eukaryotic cells. Hans Ris proposed these phrases in the 1960s. Below a microscope, you can observe that living creatures are composed of cells. Robert Hooke, a British scientist and pioneer of microscopy, coined the term "cell (Davids et al., 2010).

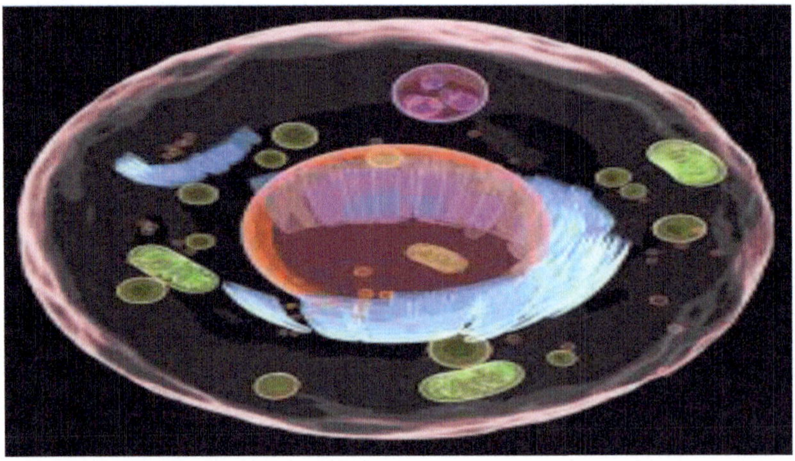

Figure 1.1. Graphical Representation of a Cell

**Source:** https://www.toppr.com/guides/biology/cell-structure-and-function/introduction-to-cell/

### 1.1.1. Animal and Plant Cell

Cells are the main components of all living things. Cells derive from different shapes and sizes, as well as different functions and transparency. Cells have metabolic capacities, which allow them to live autonomously and play an important part in living organisms. Experts and other scientists tried to figure out how cells play important roles in all living creatures, and how their lack contributes to the extinction of living beings. We now understand that there have been animal cells and plant cells, as well as these cells, that have a significant proportion of commonalities. Nevertheless, some structural distinctions have previously been described (Clément, 2007).

### 1.1.2. Features of Cells

Cells are constructed from similar molecular building blocks and have certain properties. Even though cells share several characteristics, we recognize a variety of distinct cell types; this grouping and classification of cells are referred to as cellular variety. This variety of cells varies according to the type of creature and in metazoan or multicellular life. Cells share some well-known traits, such as the use of the similar carbon atom in macromolecules, the main source of cells (Steinman & Nussenzweig, 1980). Carbohydrates, proteins, lipids, and two nucleic acids are contained inside the nucleus of live cells. Every cell contains DNA, which contains genetic information and is referred to as the genetic material of living creatures. All living species utilize their genetic material (DNA) to build proteins, which it decodes before production. Such proteins aid in the energy production of the cell, as well as its metabolic processes. Cells contain the capacity to develop and multiply even though certain cells present in multicellular animals have abandoned their capacity to differentiate, such as neurons (Damri et al., 2009).

### 1.1.3. Organelles

All animal and human cells have distinct organelles, which are connected to other cell parts through an inner layer. The nuclear membrane divides DNA from the cytoplasm, forming an essential big organelle termed the nucleus. Organelles such as mitochondria are critical organelles because they generate adenosine triphosphate (ATP), which would be a necessary part of giving vital energy for some biochemical reactions that result in the production of molecules from tiny components; an instance is the production of protein from amino acids (Dell'Angelica et al., 2000).

## 1.1.4. Building Blocks of Life

All plants and animals, including branches in a jungle, fish in a lake, horseflies on a field, lemurs in the rainforest, reeds in a canal, even insects in the earth, are formed up of the basic components we term cells. Several living organisms, such as these instances, are made up of a large cell count that functions together. Nevertheless, some kinds of life, including bacteria and protozoa, are formed entirely of a single cell. With no need for a light microscope, cells, either living alone or as a component of a multicellular, are typically too tiny to observe (Kitadai & Maruyama, 2018).

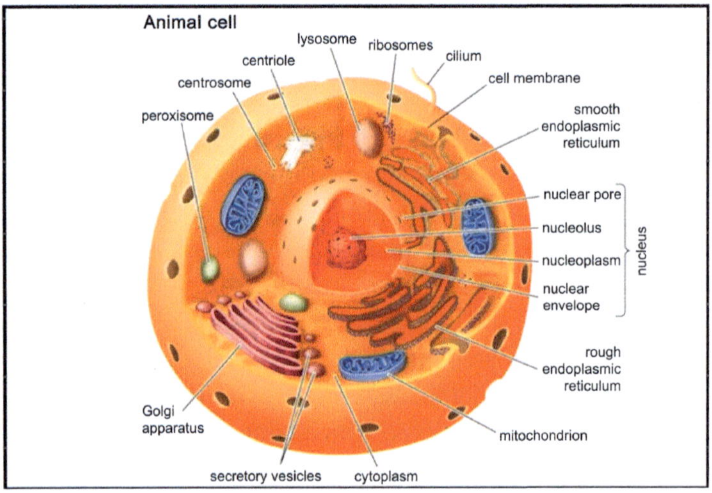


Figure 1.2. Components of a Cell

**Source:** https://www.thoughtco.com/all-about-animal-cells-373379

Cells are very much similar, but they may seem quite diverse. Cells have evolved to a broad range of surroundings and functional activities over millions and millions of years (Pirozzi et al., 2018). Nerve cells, for instance, contain lengthy, narrow extensions that may reach meters in length and are used to send impulses quickly. A tight outermost part of tightly fitting, brick-shaped plant cells enables to give the structural strength that trees and other plants need (Marth, 2008). Lengthy, tapered muscle cells contain a natural stretchy fabric that enables them to alter length when the biceps contract and relax.

Even though these cells are all dissimilar, they all use similar basic methods to retain the outer surface, improve the functioning substances while allowing everyone else to exit, manage their health, and recreate themselves. These characteristics are, in actuality, what distinguishes a cell from others (Weinkauf et al., 2000).

## 1.2. HISTORY AND ORIGIN OF CELLS

Loewy and Siekevitz described a cell as "a component of bioactivities circumscribed by a selectively permeable layer and able for self-reproduction in an environment devoid of many other biological systems" (1963). The light microscope allows the analysis of cells. Robert Hooke (1665) found that a slice of cork is composed of tiny spaces enclosed by hard walls using a light microscope (Desai & Dhanani, 2004). For the very first time, he utilized the word "cell" to characterize his observations as the "texture of a bit of cork." Later in life, A. Van Leeuwenhoek (1632-1723) examined a variety of unicellular creatures and cells, including bacteria, protozoa, red blood cells, and sperm. He discovered the nucleus in certain erythrocytes, which was become feasible by the advancements in microscopes. Marble M. declared in 1809 that almost all plant tissues are made up of cells (Stevenson et al., 1998).

J.B. Lamarck explained the relevance of organelles in living beings in a similar year. In 1831, Robert Brown discovered a nucleus in some plant cells. Dutrochet (1837) separated mimosa cells using nitric acid, concluding that all-natural tissues are made up of globular molecules held together by easy adhesive forces. Schwann, T. (1839) declared that "all living organisms are formed of cells" after analyzing a range of animal and plant tissues (Murchison et al., 2014).

The microscope's invention impacted the finding of cells. In 1665, the English microscopist and physicist Robert Hook (1635-1702) published the first explanation of cells. His scientific inquiry used thin cork slices and compared the boxy divisions he noticed to cells seen in the monastery. Hook saw wide blank areas but suggested that these openings may be utilized to carry fluid throughout a live plant. They did not verify that they were viewing the fundamental characteristics of living beings (Kubick et al., 2021).

Marcello Malpighi (1628–1694) and Hooke's colleague, Nehemiah Grew (1641–1712), proceeded to do extensive studies on plant cells and recognized the cellular structure of the plant body. Produced compared cell blank spaces to the air pockets in the risen dough and proposed that they are produced similarly. Animal cells were found later since thin sheets were required for microscope viewing and were hard to make. Today, biologists are completely sure that biological things are composed of basic components.

It piqued our interest in discovering whatever those numbers are (Mauvais-Jarvis et al., 2004). Microscopes evolved to make their observations more precise and to aid in the understanding of cells, and the microscope was selected as a critical device for studying life on the planet. In 1676, the Dutch microscopist Antony van Leeuwenhoek (1632–1723) released his study and discoveries on single-cell creatures, which he dubbed "tiny animalcules." He was a renowned scientist who was the first to view red blood cells or even sperm cells under a microscope. Leeuwenhoek found and described numerous microbes, but hundreds of years went without a link being made among cellular life and the cells which form plants and animals (Mazeh et al., 2012).

The study proceeded and culminated in 1824 when Frenchman Henri Milne-Edwards published his views on animal tissues. However according to him, animal tissues are arranged similarly to a cluster of globules (the fundamental arrangement of all animal tissues was a collection of "globules). Henri Dutrochet (1776–1847) explicitly described the link between plant to animal cells, stating as a cell is both a structural and functional unit and defining that everything originates from cells (Rossi et al., 2012).

Dutrochet suggested that new cells originate from old cells, as well as François Raspail (1794–1878), a revolutionary of Dutrochet, echoed this concept. Raspail is credited with being the primary to notice one of its two central tenets of cell theory: Omnis cellula e cellula, which translates as "Every cell is resulting from another cell" (Motta et al., 1997).

## 1.3. THE CELL THEORY

Biologists eventually revealed cells all over the place. All living creatures were proposed to be composed of cells by biologists in the early nineteenth century, however, the function of cells as the main components of life was not found till 1839, when two German experts, zoologist Theodor Schwann and botanist Matthias Jakob Schleiden, proposed, cells were the fundamental part of all living organisms (Mazzarello, 1999). Cells divide to make additional cells, as German scientist Rudolf Virchow discovered in 1858. All cells, he said, are born from other cells. The cell hypothesis is based on the findings of all three scientists... The modern cell theory was written that:

- All living things are composed of one or even more cells.
- All of an individual's life activities take place inside cells.
- All cells are derived from pre-existing ones (Baluška et al., 2004).

The cell theory, like any other theory, is derived from the work that has supported the basic findings of Schwann's 1839 paper over time. One amongst Schwann's original findings, nevertheless, was that cells were created in the same manner that crystals did. While Virchow suggested that almost all cells emerge from only those certain cells, this analysis, which also relates to the spontaneous generation of life, was dismissed. The cell theory has endured the scrutiny of modern strong microscopes and other tools used to examine cells. Scientists examine cells using new technology and tools to uncover new interpretations of how they function (Quesenberry et al., 2002).

## 1.4. CELL ARCHITECTURE

Cells are regarded as the fundamental elements of life partly since they arrive in separate and identifiable bundles. This is because all cells are enclosed by a membrane known as the cell membrane, which is similar to the walls of a home, and acts as a distinct barrier between the cell's inner and exterior surroundings. The plasma membrane is another name for the cell membrane (Herrmann et al., 2007).

As said before, the cytoplasm of a cell houses a variety of functional and structural components. These components occur in the kind of molecules and organelles; think about them as the cell's instruments, appliances, and interior chambers. Carbohydrates, nucleic acids, lipids, and proteins are all the main types of intracellular chemical compounds that are required for cell activity (Weeds, 1982).

Nucleic acids are molecules that hold and aid in the expression of a cell's genetic information. Nucleic acids are classified into two types: deoxyribonucleic acid (DNA) and ribonucleic acid (RNA). DNA is the molecule that holds all of the data necessary to construct and sustain the cell; RNA plays several functions in the translation of the data contained in DNA (O'Reilly et al., 2020).

Proteins are also used by cells to aid in the replication of the genome and the substantial structural alteration that occurs during cell division.

Proteins are a sort of intracellular organic molecule that comes in two varieties. These chemicals are composed of sequences of simpler molecules known as amino acids, and they perform a range of enzymatic and structural activities in the cell. Enzymes, for instance, are proteins that transform cellular components (such as proteins, carbohydrates, lipids, or nucleic acids) into other types that may assist a cell to satisfy its energy demands, developing support structures, or excreting waste (Hau et al., 2010). Carbohydrates, which include starches and sugars in cells, are another kind of chemical molecule. Simple carbohydrates meet the cell's direct energy requirements, while complex carbs function as intracellular energy storage. Complex carbohydrates are also present on the surface of cells, where they serve an important role in cell identification (Takeuchi et al., 1998).

Lastly, lipids or fat molecules are found in cell membranes, including the plasma membrane and several intracellular membranes. They also play a role in storing energy and signaling inside cells as well as through the circulation to the cell's interior (Herrmann et al., 2009).

Organelles, which are organized groupings of molecules in certain cells, are also present. These structures, like rooms in a home, are isolated from the main of the cell's interior through their internal membrane. Organelles include highly specialized tools essential for certain cell tasks. The mitochondrion sometimes referred to as the cell's "power plant," is one instance. It is the organelle that houses and manages the equipment complicated in energy-producing chemical processes (Warren et al., 2005).

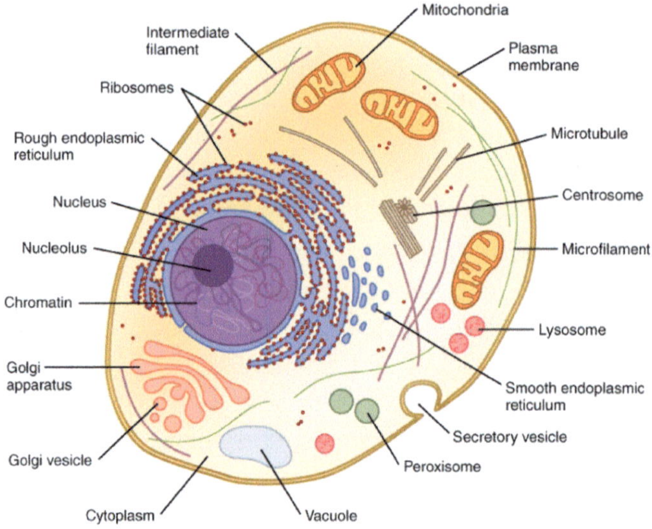


Figure 1.3. Parts of a Cell.

**Source:** https://nci-media.cancer.gov/pdq/media/images/761780.jpg

Cell membranes are made up of phospholipids, which seem to be fat-based molecules that inhibit water-loving, or hydrophilic, substances from leaving and towards the organism. Such membranes are likewise densely packed along with proteins that fulfill a variety of functions. A few of these proteins serve as gatekeepers, deciding which chemicals may and cannot pass through the membrane (Aaron et al., 2012). Others serve as identifiers, distinguishing the cell as belonging to a similar organism or as an alien. Some, such as fasteners, connect cells so that they may operate as a whole. Some membrane proteins, on the opposite side, act as communicators, transmitting and receiving messages from nearby cells and the environment, either friendly or frightening.

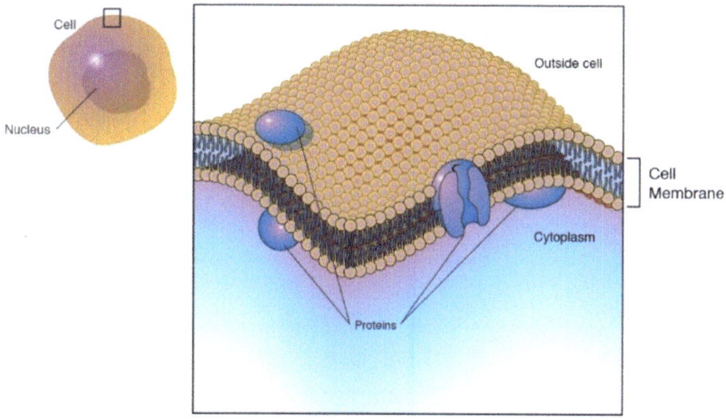


Figure 1.4. Architecture of a Cell.

**Source:** https://www.genome.gov/sites/default/files/tg/en/illustration/plasma_membrane.jpg

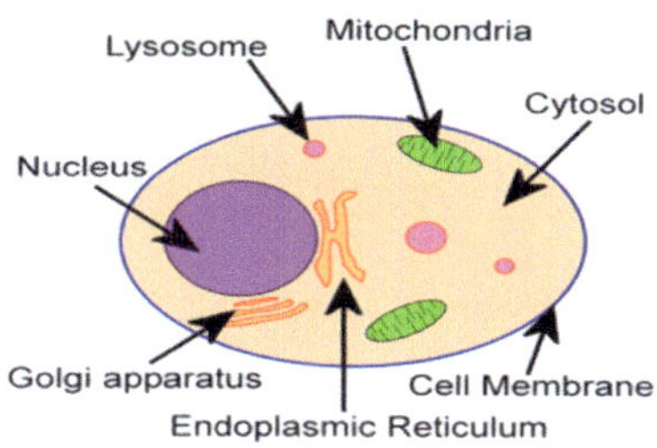


Figure 1.5. Components of Cells.

**Source:** http://www.timkelf.com/research/ResearchCellBio.html

The internal environment of a cell is water-based inside this membrane. This liquid environment, known as cytoplasm, is tightly packed with cellular receptors and structural parts (Kaiser et al., 2008). In reality, protein concentrations within a cell much outweigh the others on the exterior, if the exterior is ocean water (like in the scenario of a single-celled alga) or blood serum (like in the scenario of a red blood cell). Even though cell membranes constitute natural borders in watery environments, a cell needs to waste a significant amount of energy to sustain the large levels of internal elements required for life. However, cells may utilize up to 30% of their energy simply to retain the makeup of their cytoplasm (Schroeder & Halder, 2012).

### 1.4.1. Cell Nucleus

It is an essential component of the cell that was first mentioned in 1802 by a Czech, Franz Bauer, and was titled in 1831 by Robert Brown (1773–1858) of Scotland, and has since been described in these other sections of the nucleus. In 1838 and 1839, Schleiden and Schwann summarised their studies on cell biology and highlighted their contributions (Dundr & Misteli, 2001). Matthias Schleiden (1804–1881) made clear his theory in 1838, stating that all plant kinds or components are composed of cells. In 1839, a fellow German named Theodor Schwann (1810–1882) proposed ideas about animal structure. All structural components in animals, he claimed, are cell set brands, meaning they are created by cells. Schwann's contribution appears to be emulating which cell theory on plants has been recommended. He claimed that the rules regulating cells in plants and animals were similar or identical (Görlich & Kutay, 1999).

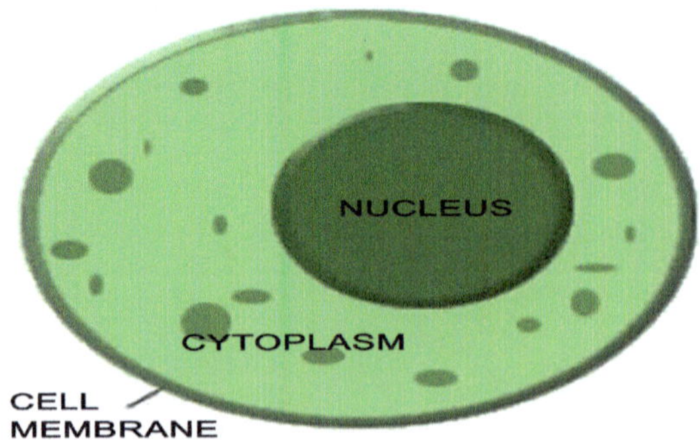


Figure 1.6. Nucleus of a Cell.

Source: https://jrobinsonjparletteorganelles.weebly.com/basic-cell-components.html

Jan Purkye (1787–1869), often referred to as Purkinje, was a Czech singular cytologist and among the most fundamental synthetic chemists of cell theory. He explained his work using Schwann's theory. His theory though is that animals are made up of cells and cell products and that the same theory could be extended to plants. Many scientists participated in cell theory as well, and those are the most important (Guilak et al., 2000).

### 1.4.2. Unity of Cells

Cell biologists frequently refer to "the cell" without naming any specific cell. Cells, on the other hand, may not all be alike; in reality, they could be quite dissimilar. According to biologists, our planet could have up to 100 million different organisms of living things. We must first take inventory before digging further into cell biology: What else do a bacteria and a butterfly have similar? What do the cells of a rose and a dolphin have similarities (Celma et al., 1995)?

And how does the diversity of cell types inside a single multicellular creature vary?

The Appearance and Function of Cells Vary Extremely Let's start with the size. A bacterial cell is several micrometers (m) long, such as a Lactobacillus in a slice of cheese. This is roughly a quarter of the breadth of human hair. The diameter of a frog egg, which would be likewise a unicellular organism, is roughly 1 millimeter. The frog egg would've been half a mile high if we sized it ahead to create the Lactobacillus the size of a human (Alzaher & Ismail, 1999).

The morphology of cells varies exactly as much. A normal nerve cell in your brain, for instance, is very lengthy; it gives out electrical impulses with a fine protrusion 10,000 times larger as compared to its thickness, and it gets data from many other nerve cells via a myriad of smaller procedures which sprout out of its center as tree branches (Toker et al., 2001). A paramecium in pond water has the form of a submarine and is coated in hundreds of fimbriae expansions whose sinuous pulsing propels the cell ahead, spinning as it does so. A squat and stationary cell in a plant's upper layers is encased in a stiff tube of cellulose with only an exterior waterproof covering of wax. In the animal body, however, a neutrophil or macrophage crawls through the tissues, continuously reshaping itself because it seeks for as well as envelops debris, foreign microbes, dead or injured cells, and so on (TaherAbuelma'atti & AliAl-Qahtani, 1996).

The chemical needs of cells are likewise quite different. Many people need oxygen to survive, while others are poisoned by it. A few cells require only air, sunlight, and water as rare resources, whereas others require a complicated set of products formed by other cells. Cell function is typically reflected in changes in size, structure, and metabolic needs. Many cells function as specialized manufacturers for the synthesis of specific substances like hormones, starch, fat, latex, and pigments. Some, such as muscle cells, are engines that consume energy to do physiological labor (Chang et al., 2001). Others, such as the altered muscle tissue in the electric eel, are electricity producers. Certain alterations make a cell so specialized that this loses its ability to leave offspring. For a cellular that existed alone, this specialization would be pointless. However, in a complex organism, there is a separation of work between cells, enabling certain cells into becoming highly specialized for certain functions while remaining reliant on their partner cells for many fundamental needs. This most fundamental requirement of everyone, the transmission of the organism's genetic instructions to the next cycle, is given to professionals' eggs and sperm (Gupta & Senani, 2006).

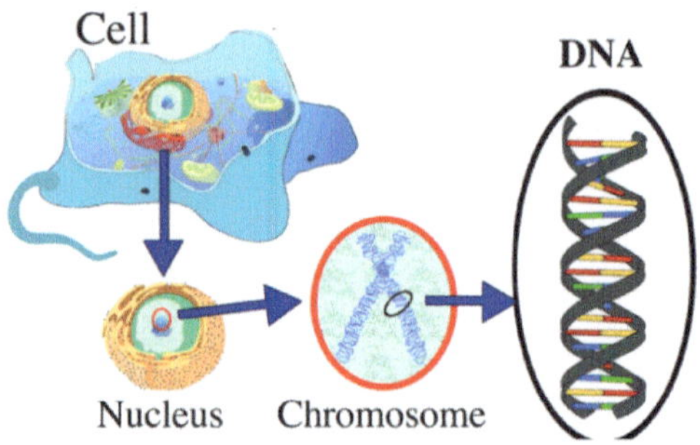


Figure 1.7. Units of a Cell.

**Source:** https://newsela.com/read/natgeo-cell-discovery-history/id/50290/

### 1.4.3. Diversity of Cells

Inside the same species, cells may vary in size and form. They might not have been large, however, their forms might be rather varied. Nevertheless, all of these cells have some capabilities, including obtaining as well as utilizing energy from food, reacting to their surrounding environment, and multiplying. The form of a cell dictates its function (Celma et al., 1995).

### 1.4.4. Cell Size

Why are cells so tiny if they play a crucial role? Why aren't there any animals along with colossal cells? The replies to such questions may be determined by a cell's wish for rapid, easy meals. They need to be capable of moving nourishment and gas to and from the cell boundaries the size of cells. That when a cell raises, harder it develops for nourishment and gases to go in and out (Lloyd, 2013).

A cell's volume expands faster as compared to the surface area as it develops. The limited surface area of a cell just wouldn't enable sufficient nourishment to reach the cell rapidly sufficient to meet the cell's demands if it grew to be extremely big. This concept is discussed in depth in this section. Large cells, on the other hand, have a manner of coping with certain size constraints (Ginzberg et al., 2015).

Big cells, like certain white blood cells, often generate additional nuclei to deliver adequate proteins and RNA for the cell's requirements. The cell surface membrane of big, metabolically active cells frequently has several folds. The surface accessible for transit to or from the cell is increased by such folds. These types of cells line the small intestine, wherein they digest food from your meals via microvilli, which are tiny folds (Stocker & Hafen, 2000).

## Scale of Measurements

1 centimeter (cm) = 10 millimeters (mm) = $10^{-2}$ meters (m)

1 mm = 1000 micrometers (µm) = $10^{-3}$ m

1 µm = 1000 nanometers (nm) = $10^{-6}$ m

1 nm = 10-3 µm

Imagine cells as little cube blocks. A small cube cell is one unit in length.

The total surface area of this cell is calculated by the equation:

height × width × number of sides × number of boxes

$1 \times 1 \times 6 \times 1 = 6$

The volume of the cell is calculated:

height x width x length x number of boxes

$1 \times 1 \times 1 \times 1 = 1$

The surface-area to volume ratio is:

area ÷ volume

$6 \div 1=6$

A larger cell that is 3 units in length would have a total surface area of

$3 \times 3 \times 6 \times 1 = 54$

and a volume of:

$3 \times 3 \times 3 \times 1 = 27$

The surface area to volume ratio of the large cell is (Taheri-Araghi et al., 2015):

$54 \div 27 = 2$

Now, replace the three-unit cell with enough one-unit cells to equal the volume of the single three-unit cell.

This can be done with 27 one-unit cells. Find the total surface area of the 27 cells:

$1 \times 1 \times 6 \times 27 = 162$ units

The total volume of the block of 27 cells is:

$1 \times 1 \times 1 \times 27 = 27$

The surface area to volume ratio of the 27 cells is:

$162 \div 27 = 6$

Higher surface-to-volume relation equates to rising environmental contact. This implies that food and gas can pass more quickly to and from a tiny cell than they can via a bigger cell. The lightest prokaryotic cell has a diameter of 400 nm. Eukaryotic cells typically have a diameter of between 1 and 100 micrometers (Nurse, 1975).

### 1.1.5 Cell Shape

The roles of every cell are reflected in the range of cell structures found in prokaryotes and eukaryotes. Every type of cell has developed a morphology that best assists it in surviving and performing its function (Thery & Bornens, 2006). Figure 1.8 shows an example of a nerve cell with lengthy, narrow projections that stretch out to neighboring nerve cells. The projections aid the nerve cell in swiftly transmitting electrical and chemical impulses throughout the body. The pollen grain's spike aid in its attachment to a pollinating bug or mammal, allowing it to be carried to and fertilize other flowers. The algae Chlamydomonas can move to its long whip-like flagella (tails) (Folkman & Moscona, 1978).

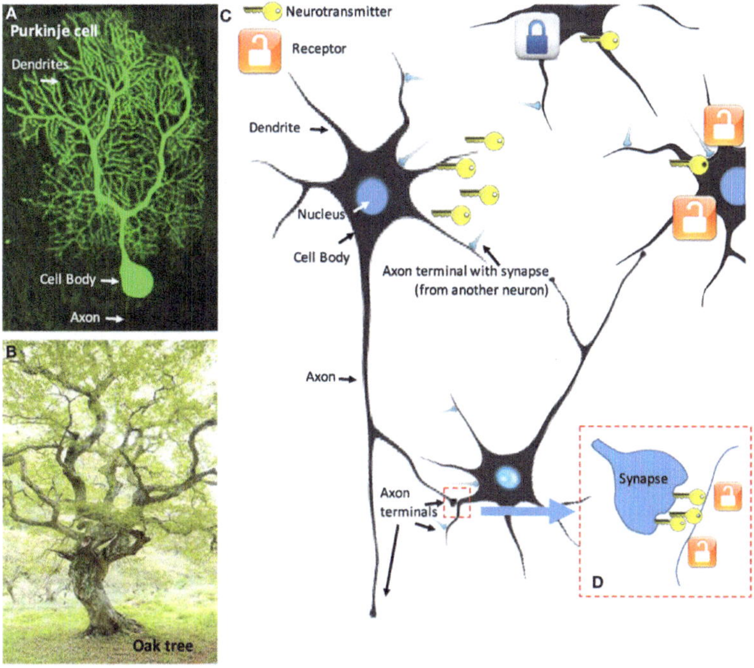


**Figure 1.8.** A typical nerve cell

**Source:** Ludwig M (2017) How Your Brain Cells Talk to Each Other—Whispered Secrets and Public Announcements. Front. Young Minds. 5:39. doi: 10.3389/frym.2017.00039

A. Some neurons look very similar to trees B. C. Neurotransmitters (key) from the axon terminals cross a very tiny gap (called a synapse) D. to reach their receptors (lock). Purkinje cell image courtesy of Marta Jelitai, Hungary

## 1.5 CHEMISTRY OF CELLS

Instead of the great variety of flora and fauna, humans have long understood that these species have a basic trait that qualifies everything as living beings. However, although it was simple to discern life, it was incredibly hard to determine how all living creatures were identical. Textbooks have to make do with theoretical generalizations about development, fertility, and the capacity to adapt to the environment (Hober et al., 1946).

Biochemists and molecular biologists' findings have supplied a graceful answer to this perplexing dilemma. While all living creatures' cells seem to be immensely different from the surface, they are essentially identical on the inside. We now understand that cells are astonishingly similar in terms of their chemistry. They are both made up of similar kinds of molecules and undergo similar types of chemical processes. In all species, genetic data is transmitted in DNA molecules in the type of genes (Blake et al., 1999).

This data is contained in the very same chemical sequence, manufactured using similar chemical basic components, processed by basically a similar chemical apparatus, and copied similarly when an organism replicates. Therefore, in each cell, the lengthy DNA polymer chains are constructed from a similar set of four monomers called nucleotides that are linked collectively in varied sequences to communicate information, much like the letters of an alphabet (Zhu et al., 2005).

Each cell reads out or transcribes, the information recorded in DNA into a chemically similar collection of polymers termed RNA. A portion of such RNA molecules is then translated into various sorts of polymer known as a protein. This data transfer—from DNA to RNA protein—is so critical to life that one is known as the core dogma (Owicki & Parce, 1992).

A cell's appearance and activity are primarily determined by its protein molecules, which function as structural supports, chemical enzymes, and molecular engines, among other functions. Proteins are constructed from amino acids, and all species employ the very same set of twenty amino acids (Hong et al., 2010).

However, amino acids are connected in a variety of ways, providing every kind of protein molecule a unique three-dimensional structure, or conformation, much as various letter patterns spell out distinct words (Rostam et al., 2015).

Thus, a similar fundamental biochemical mechanism has been responsible for the emergence of all living things on earth. The functioning of proteins, RNA, and DNA is discussed in more depth. If cells are the basic component of living stuff, therefore anything but a cell can be considered genuinely alive (Ahamed et al., 2008). Viruses, for instance, are little bundles of genetic material wrapped in protein that contain genetic information in the type of DNA or RNA, but they cannot replicate themselves. Rather than that, they replicate themselves by despoiling the reproduction apparatus of the cells they infect. Therefore, viruses are chemical monsters: they are dormant and inert outside of their host tissue, however once inside, they may exercise malignant influence over the cell (Burke & Lipomi, 2013).

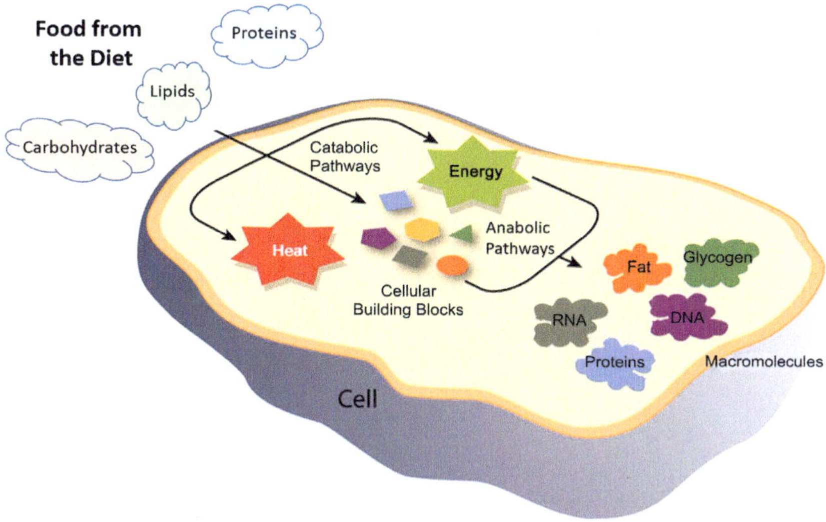


**Figure 1.9.** Catabolic reactions involve the breakdown of molecules into smaller components, whereas anabolic reactions build larger molecules from smaller molecules. Catabolic reactions usually release energy whereas anabolic processes usually require energy.

**Source:** https://wou.edu/chemistry/courses/online-chemistry-textbooks/ch450-and-ch451-biochemistry-defining-life-at-the-molecular-level/chapter-1-the-foundations-of-biochemistry/

## 1.6 EVOLUTION OF CELLS

A cell creates by copying its DNA and afterward splitting into two daughter cells, each one receives a set of genetic information recorded in its DNA. And this is why daughter cells look like their parents. Nevertheless, copying isn't always exact, and directions may be tampered with by DNA mutations (Hsu, 1962). As a result, daughter cells may not always perfectly resemble the parent cell. Mutations can result in children that are altered into something worse (in terms of their ability to live and thrive), altered for the better (in terms of their ability to successfully reproduce), or altered in a general sense (in terms of their ability to survive and reproduce) (in that they differ genetically but equally viable) (Arendt et al., 2016).

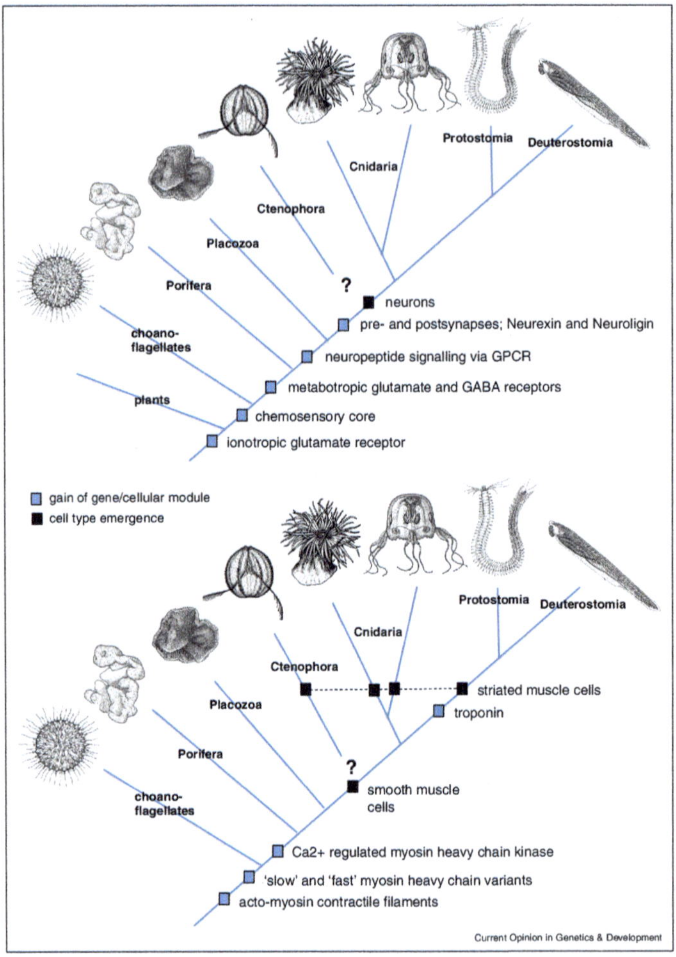


Figure 1.10. Evolution of Cells.

**Source:** https://ars.els-cdn.com/content/image/1-s2.0-S0959437X14000483-gr1.jpg

The battle for existence removes the first, favors the second, but allows the third to survive. The genes of the surviving will be passed on to the following generation. Sexual reproduction, wherein two cells of a similar species unite and pool their DNA, may often confuse the descent sequence (Arendt, 2008). The genetic cards then are mixed, re-dealt, and handed to the following generation in different formations, where they will be assessed for their potential to improve growth and survival once again. The method through which life forms are becoming progressively amended and adjusted to their surroundings in even more sophisticated forms is based on these simple rules of genetic mutation and choice, which have been applied again and again over hundreds of millions of cell generations (Hynes & Zhao, 2000). Evolution provides a strange but plausible description for why today's cells are so fundamentally the same: they all received their genetic information from a similar common ancestor. *That ancestral cell is thought to have lived around 3.5 and 3.8 billion years ago, and therefore we have to assume that it had a precursor of the global machinery that all earth-life life uses today. The offspring of this primordial cell have progressively diversified to occupy each environment on Earth with species that use the machinery's potential in an alternative manner, thanks to a very lengthy procedure of mutation as well as natural selection* (Fuerst, 1995).

## REFERENCES

1. Aaron, D. S., Liu, Q., Tang, Z., Grim, G. M., Papandrew, A. B., Turhan, A., ... & Mench, M. M. (2012). Dramatic performance gains in vanadium redox flow batteries through modified cell architecture. Journal of Power sources, 206, 450-453.
2. Ahamed, M., Karns, M., Goodson, M., Rowe, J., Hussain, S. M., Schlager, J. J., & Hong, Y. (2008). DNA damage response to different surface chemistry of silver nanoparticles in mammalian cells. Toxicology and applied pharmacology, 233(3), 404-410.
3. Alzaher, H. A., & Ismail, M. (1999). Current-mode universal filter using unity gain cells. Electronics Letters, 35(25), 2198-2200.
4. Arendt, D. (2008). The evolution of cell types in animals: emerging principles from molecular studies. Nature Reviews Genetics, 9(11), 868-882.
5. Arendt, D., Musser, J. M., Baker, C. V., Bergman, A., Cepko, C., Erwin, D. H., ... & Wagner, G. P. (2016). The origin and evolution of cell types. Nature Reviews Genetics, 17(12), 744-757.
6. Baluška, F., Volkmann, D., & Barlow, P. W. (2004). Eukaryotic cells and their cell bodies: cell theory revised. Annals of Botany, 94(1), 9-32.
7. Blake, D. M., Maness, P. C., Huang, Z., Wolfrum, E. J., Huang, J., & Jacoby, W. A. (1999). Application of the photocatalytic chemistry of titanium dioxide to disinfection and the killing of cancer cells. Separation and purification methods, 28(1), 1-50.
8. Burke, D. J., & Lipomi, D. J. (2013). Green chemistry for organic solar cells. Energy & Environmental Science, 6(7), 2053-2066.
9. Celma, S., Sabadell, J., & Martinez, P. (1995). Universal filter using unity-gain cells. Electronics Letters, 31(21), 1817-1818.
10. Chang, C. M., Liao, T. S., Yu, T. Y., Lin, E. S., Teng, C. H., & Hou, C. L. (2001). Novel universal current-mode filters using unity-gain cells. International journal of electronics, 88(1), 23-30.
11. Clément, P. (2007). Introducing the cell concept with both animal and plant cells: A historical and didactic approach. Science & Education, 16(3), 423-440.
12. Damri, M., Granot, G., Ben-Meir, H., Avivi, Y., Plaschkes, I., Chalifa-Caspi, V., ... & Grafi, G. (2009). Senescing cells share common features with dedifferentiating cells. Rejuvenation research, 12(6), 435-443.

13. Davids, M., Forrest, R., & Pata, D. (2010). Teaching the fundamentals of cell phones and wireless communications. The Physics Teacher, 48(4), 217-221.
14. Dell'Angelica, E. C., Mullins, C., Caplan, S., & Bonifacino, J. S. (2000). Lysosome-related organelles. The FASEB Journal, 14(10), 1265-1278.
15. Desai, D. V., & Dhanani, H. (2004). Sickle cell disease: history and origin. The internet journal of hematology, 1(2), 1540.
16. Dundr, M., & Misteli, T. (2001). Functional architecture in the cell nucleus. Biochemical Journal, 356(2), 297-310.
17. Folkman, J., & Moscona, A. (1978). Role of cell shape in growth control. Nature, 273(5661), 345-349.
18. Fuerst, J. A. (1995). The planctomycetes: emerging models for microbial ecology, evolution and cell biology. Microbiology, 141(7), 1493-1506.
19. Ginzberg, M. B., Kafri, R., & Kirschner, M. (2015). On being the right (cell) size. Science, 348(6236), 1245075.
20. Görlich, D., & Kutay, U. (1999). Transport between the cell nucleus and the cytoplasm. Annual review of cell and developmental biology, 15(1), 607-660.
21. Guilak, F., Tedrow, J. R., & Burgkart, R. (2000). Viscoelastic properties of the cell nucleus. Biochemical and biophysical research communications, 269(3), 781-786.
22. Gupta, S. S., & Senani, R. (2006). New single-resistance-controlled oscillator configurations using unity-gain cells. Analog integrated circuits and signal processing, 46(2), 111-119.
23. Hau, S. K., Yip, H. L., & Jen, A. K. Y. (2010). A review on the development of the inverted polymer solar cell architecture. Polymer Reviews, 50(4), 474-510.
24. Herrmann, H., Bär, H., Kreplak, L., Strelkov, S. V., & Aebi, U. (2007). Intermediate filaments: from cell architecture to nanomechanics. Nature reviews Molecular cell biology, 8(7), 562-573.
25. Herrmann, H., Strelkov, S. V., Burkhard, P., & Aebi, U. (2009). Intermediate filaments: primary determinants of cell architecture and plasticity. The Journal of clinical investigation, 119(7), 1772-1783.
26. Hober, R., Hitchcock, D. I., Bateman, J. B., Goddard, D. R., & Fenn,

W. O. (1946). Physical Chemistry of Cells and Tissues. The Journal of Physical Chemistry, 50(4), 386-387.

27. Hong, V., Steinmetz, N. F., Manchester, M., & Finn, M. G. (2010). Labeling live cells by copper-catalyzed alkyne– azide click chemistry. Bioconjugate chemistry, 21(10), 1912-1916.
28. Hsu, T. C. (1962). Chromosomal evolution in cell populations. In International review of cytology (Vol. 12, pp. 69-161). Academic Press.
29. Hynes, R. O., & Zhao, Q. (2000). The evolution of cell adhesion. The Journal of cell biology, 150(2), F89-F96.
30. Kaiser, J. P., Wick, P., Manser, P., Spohn, P., & Bruinink, A. (2008). Single walled carbon nanotubes (SWCNT) affect cell physiology and cell architecture. Journal of Materials Science: Materials in Medicine, 19(4), 1523-1527.
31. Kitadai, N., & Maruyama, S. (2018). Origins of building blocks of life: A review. Geoscience Frontiers, 9(4), 1117-1153.
32. Kubick, N., Klimovich, P., Flournoy, P. H., Bieńkowska, I., Łazarczyk, M., Sacharczuk, M., ... & Basu, R. (2021). Interleukins and interleukin receptors evolutionary history and origin in relation to CD4+ T cell evolution. Genes, 12(6), 813.
33. Lloyd, A. C. (2013). The regulation of cell size. Cell, 154(6), 1194-1205.
34. Marth, J. D. (2008). A unified vision of the building blocks of life. Nature cell biology, 10(9), 1015-1015.
35. Mauvais-Jarvis, F., Sobngwi, E., Porcher, R., Riveline, J. P., Kevorkian, J. P., Vaisse, C., ... & Gautier, J. F. (2004). Ketosis-prone type 2 diabetes in patients of sub-Saharan African origin: clinical pathophysiology and natural history of β-cell dysfunction and insulin resistance. Diabetes, 53(3), 645-653.
36. Mazeh, H., Benavidez, J., Poehls, J. L., Youngwirth, L., Chen, H., & Sippel, R. S. (2012). In patients with thyroid cancer of follicular cell origin, a family history of nonmedullary thyroid cancer in one first-degree relative is associated with more aggressive disease. Thyroid, 22(1), 3-8.
37. Mazzarello, P. (1999). A unifying concept: the history of cell theory. Nature cell biology, 1(1), E13-E15.
38. Motta, P. M., Nottola, S. A., & Makabe, S. (1997). Natural history of the female germ cell from its origin to full maturation through prenatal

ovarian development. European Journal of Obstetrics & Gynecology and Reproductive Biology, 75(1), 5-10.

39. Murchison, E. P., Wedge, D. C., Alexandrov, L. B., Fu, B., Martincorena, I., Ning, Z., ... & Stratton, M. R. (2014). Transmissible dog cancer genome reveals the origin and history of an ancient cell lineage. Science, 343(6169), 437-440.
40. Nurse, P. (1975). Genetic control of cell size at cell division in yeast. Nature, 256(5518), 547-551.
41. O'Reilly, F. J., Xue, L., Graziadei, A., Sinn, L., Lenz, S., Tegunov, D., ... & Rappsilber, J. (2020). In-cell architecture of an actively transcribing-translating expressome. Science, 369(6503), 554-557.
42. Oshima, J., & Campisi, J. (1991). Fundamentals of cell proliferation: control of the cell cycle. Journal of dairy science, 74(8), 2778-2787.
43. Owicki, J. C., & Parce, J. W. (1992). Biosensors based on the energy metabolism of living cells: the physical chemistry and cell biology of extracellular acidification. Biosensors and Bioelectronics, 7(4), 255-272.
44. Pirozzi, N. M., Hoogenboom, J. P., & Giepmans, B. N. (2018). ColorEM: analytical electron microscopy for element-guided identification and imaging of the building blocks of life. Histochemistry and cell biology, 150(5), 509-520.
45. Quesenberry, P. J., Colvin, G. A., & Lambert, J. F. (2002). The chiaroscuro stem cell: a unified stem cell theory. Blood, The Journal of the American Society of Hematology, 100(13), 4266-4271.
46. Rossi, D., Spina, V., Forconi, F., Capello, D., Fangazio, M., Rasi, S., ... & Gaidano, G. (2012). Molecular history of Richter syndrome: origin from a cell already present at the time of chronic lymphocytic leukemia diagnosis. International journal of cancer, 130(12), 3006-3010.
47. Rostam, H. M., Singh, S., Vrana, N. E., Alexander, M. R., & Ghaemmaghami, A. M. (2015). Impact of surface chemistry and topography on the function of antigen presenting cells. Biomaterials science, 3(3), 424-441.
48. Schroeder, M. C., & Halder, G. (2012, September). Regulation of the Hippo pathway by cell architecture and mechanical signals. In Seminars in cell & developmental biology (Vol. 23, No. 7, pp. 803-811). Academic Press.
49. Sharaf, O. Z., & Orhan, M. F. (2014). An overview of fuel cell

technology: Fundamentals and applications. Renewable and sustainable energy reviews, 32(1), 810-853.

50. Steinman, R. M., & Nussenzweig, M. C. (1980). Dendritic cells: features and functions. Immunological reviews, 53(1), 127-147.
51. Stevenson, F., Sahota, S., Zhu, D., Ottensmeier, C., Chapman, C., Oscier, D., & Hamblin, T. (1998). Insight into the origin and clonal history of B-cell tumors as revealed by analysis of immunoglobulin variable region genes. Immunological Reviews, 162(1), 247-259.
52. Stocker, H., & Hafen, E. (2000). Genetic control of cell size. Current opinion in genetics & development, 10(5), 529-535.
53. TaherAbuelma'atti, M., & AliAl-Qahtani, M. (1996). Current-mode universal filters using unity gain cells. Electronics Letters, 32(12), 1077-1079.
54. Taheri-Araghi, S., Bradde, S., Sauls, J. T., Hill, N. S., Levin, P. A., Paulsson, J., ... & Jun, S. (2015). Cell-size control and homeostasis in bacteria. Current Biology, 25(3), 385-391.
55. Takeuchi, K., Tanaka, T., & Tanzawa, T. (1998). A multipage cell architecture for high-speed programming multilevel NAND flash memories. IEEE Journal of Solid-State Circuits, 33(8), 1228-1238.
56. Thery, M., & Bornens, M. (2006). Cell shape and cell division. Current opinion in cell biology, 18(6), 648-657.
57. Toker, A., Zeki, A., & Cicekoglu, O. (2001). Generalized gyrator implementation techniques using unity-gain cells. International journal of electronics, 88(5), 517-528.
58. Warren, D. T., Zhang, Q., Weissberg, P. L., & Shanahan, C. M. (2005). Nesprins: intracellular scaffolds that maintain cell architecture and coordinate cell function?. Expert reviews in molecular medicine, 7(11), 1-15.
59. Weeds, A. (1982). Actin-binding proteins—regulators of cell architecture and motility. Nature, 296(5860), 811-816.
60. Weinkauf, R., Schermann, J. P., De Vries, M. S., & Kleinermanns, K. (2002). Molecular physics of building blocks of life under isolated or defined conditions. The European Physical Journal D-Atomic, Molecular, Optical and Plasma Physics, 20(3), 309-316.
61. Zhu, H., Kee, R. J., Janardhanan, V. M., Deutschmann, O., & Goodwin, D. G. (2005). Modeling elementary heterogeneous chemistry and electrochemistry in solid-oxide fuel cells. Journal of the electrochemical society, 152(12), A2427.

# CHAPTER 2

# STRUCTURE OF CELL

CONTENTS

## 2.1 INTRODUCTION

Over time, people's perceptions of cell structure have shifted dramatically. According to early scientists, cells were formerly assumed to be basic membrane sacs that contained fluid and a few floating particles. Cells, according to today's experts, have been far more complex than this throughout their evolution (Chothia & Jones, 1997). The cytoplasm and nucleus are two parts of the cell's interior. The nucleus is an elliptic or spherical structure found at the cell's center. Outside of the nucleus, the cytoplasm comprises cytosol and cell organelles, or cytoplasmic solution. The fluid and the cytosol inside the nucleus and organelles make up intracellular fluid (Fischer et al., 2003).

There are 3 parts of the cell, which are:

- Cytoplasm
- Nucleus
- Cell Membrane

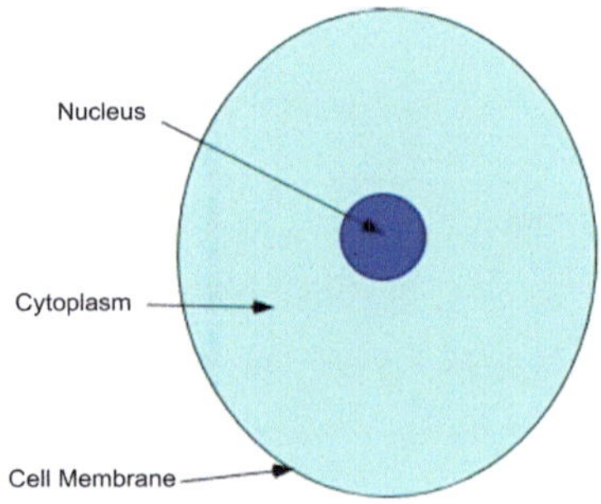


Figure 2.1. Parts of a Cell.

**Source:** https://breast-cancer.ca/cytoplagarsm/

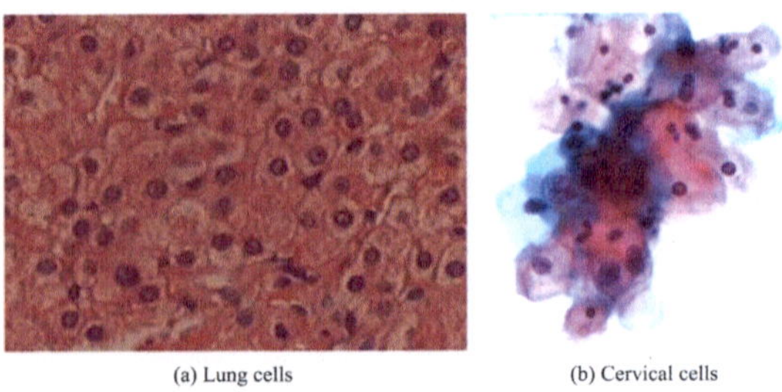


Figure 2.2. Cell Structure.

**Source:** Chen, S., Zhao, M., Wu, G., Yao, C., & Zhang, J. (2012). Recent advances in morphological cell image analysis. Computational and mathematical methods in medicine, 2012, 101536. https://doi.org/10.1155/2012/101536

## 2.2. CELL MEMBRANE

Each living cell is surrounded by a thin membrane that is normally between four and ten nanometers (nm; One nm = 109 metres) thick, separating it from its surroundings. The cell membrane contains the cell's components, which are usually large, water-soluble, highly charged molecules such as carbohydrates, nucleic acids, proteins, and substances involved in cell metabolism (also referred to as the plasma membrane) (Yeagle, 1985). The surrounding water-based environment contains alkalis, acids, and ions that are damaging to the cell, as well as nutrients that the cell needs to live and thrive. As a consequence, the cell membrane performs two functions: first, like a barrier that keeps cell components in while keeping unwanted items out, and second, as a gate that enables essential nutrients into the cell while allowing waste products to depart (Mohandas & Gallagher, 2008).

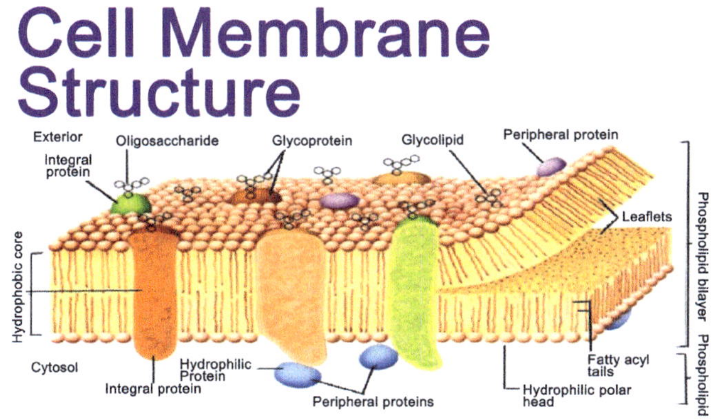


Figure 2.3. Membrane Molecular Structure.

**Source:** https://slideplayer.com/slide/14451036/

### 2.2.1. Chemical Composition and Membrane Structure

It is from studies of red blood cells that we have derived the majority of our present research on the biochemical elements of cell membranes. The primary benefit of using such cells for experimental purposes is that they have been inexpensive and easy to produce in great quantities, and they do not have any inner membranous organelles which can interfere with the study of the membranes of their cells (Branton & Deamer, 1972). In-depth examinations of this and other cell types have revealed that certain membranes are made of fatty acid lipids and proteins, except for plasma membranes. Actively metabolic membranes comprise a high proportion of protein; for example, the mitochondrion membrane, the cell's most quickly metabolizing organelle, encompasses up to seventy-five percent protein, whereas the membrane of the Schwann cell, that forms an insulation sheath around several nerve cells, includes minimum twenty percent protein (Parpart & Dziemian, 1940).

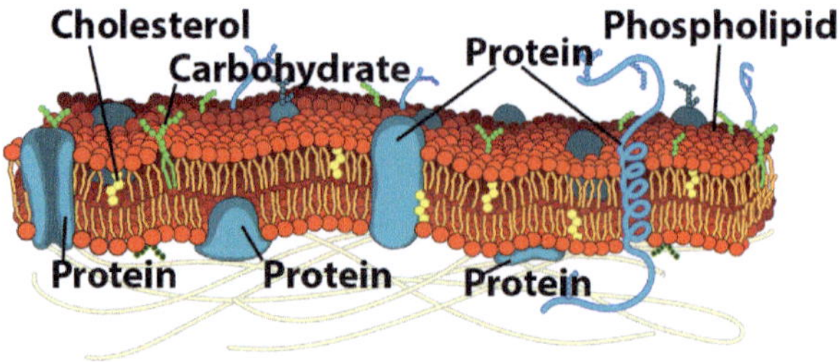


Figure 2.4. Cell Membrane Composition.

**Source:** https://learn-biology.com/membranes-2-membrane-structure/

## 2.2.2. Membrane Lipids

Sterols and phospholipids are the two types of membrane lipids (typically cholesterol). Both variants have an aqueous-attractive region that dissolves in aqueous, which is a distinguishing property of lipids: they break down swiftly in organic solvents. The "amphiphilic" property of lipids determines their relevance as a component of cell membranes (bearing a dual attraction; that is, including both a water-soluble and a lipid-soluble area) (Van Meer et al., 2008). Phospholipid molecules have a head (usually formed of glycerol) to which two long fatty acid chains that resemble tails are connected. Water repels such tails, which dissolve quickly in organic solvents and confer lipid properties to the molecule. A phosphoryl group with a -ve electrical charge has been connected to another area of the head; this group, in turn, has been related to a +ve or neutral charge group. The amphiphilic character of the molecule is established when this region of the phospholipid disintegrates in water. Sterols have a lipid-soluble segment with a complex hydrocarbon ring structure on one side and a water-soluble region with a hydroxyl grouping on the other (Cronan Jr et al., 1975).

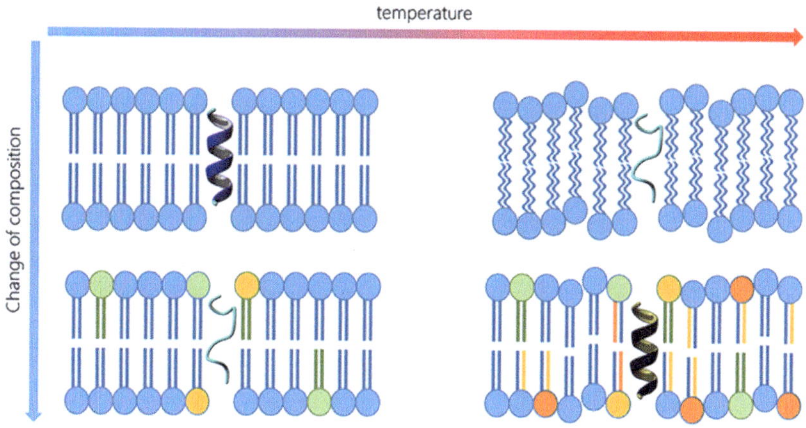


Figure 2.5. Membrane Lipids.

**Source:** Piotto S, Di Biasi L, Sessa L and Concilio S (2018) Transmembrane Peptides as Sensors of the Membrane Physical State. Front. Phys. 6:48. doi: 10.3389/fphy.2018.00048

Liposomes are globular structures formed whenever dry phosphor-lipids, or a combination of this cholesterol and phosphor-lipids, are submerged in water under controlled circumstances. Liposomes are built up of concentric spheres, one within the other, every constituting half of a two-layered wall, according to research. A two-layer phospholipid molecule is made up of two sheets, each with all of the molecules facing an identical direction (Sprott, 1992). In an aqueous medium, the water-repellent, lipid-soluble tails of one sheet's molecules are twisted and insecurely attached to the tails of the molecules on the other sheet. The water-soluble heads spin outward into the water and are attracted to it chemically. In this way, sandwich-like structure, the two sheets form a fluid, with the fatty acid chains in the center mixing in an organic medium while the watery media is shut off (Ernst et al., 2016).

The fundamental structure of the cell membrane is a lipid bilayer generated by the self-assembly of lipid molecules (Sun, G. & Sun, A., 1985). The water-attractive region covered in water and pointing away from the fatty acid zones was the most thermodynamically stable structure a phospholipid-water mixture could adopt in the phase of organically formed by comparable other molecules regions, and the water-attractive region covered in water and pointing away from the fatty acid zones was the most thermodynamically steady structure a phospholipid-water mixture could adopt. The bilayer structure, therefore, fulfills every region of the amphiphilic molecule's chemical attraction (Cronan, 2003).

### 2.2.3. Membrane Proteins

Membrane proteins have been divided into 2 categories. Extrinsic proteins are loosely associated with the electrically charged phosphoryl surface of the bilayer via ionic connections or calcium bridges. They may also bind to intrinsic proteins, which are the 2nd kind of protein. Intrinsic proteins have been securely anchored inside the phospholipid bilayer as its name indicates. Almost all intrinsic proteins have unique amino acid sequences which extend into the cell membrane's interior regions, usually between twenty and twenty-four amino acids long (Guidotti, 1972).

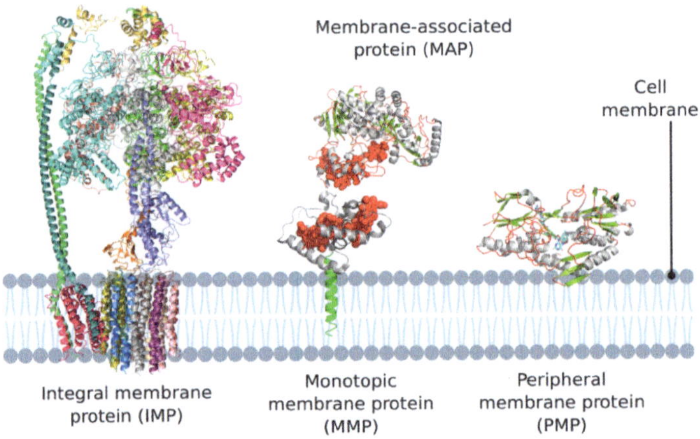


**Figure 2.6.** Membrane protein structural categorization.

**Source:** Boes, D. M., Godoy-Hernandez, A., & McMillan, D. (2021). Peripheral Membrane Proteins: Promising Therapeutic Targets across Domains of Life. *Membranes*, *11*(5), 346. https://doi.org/10.3390/membranes11050346

Several extrinsic and intrinsic proteins have side chains of complex sugars on their external surface that stretch into the aqueous environment outside the cell. Glycoproteins are the name given to such proteins because of this. Glycoproteins have a role in cell-to-cell communication (Bowie, 2001).

### 2.2.4. The Fluidity of the Membrane

The discovery of the cell membrane like a fluid accumulation of amphiphilic molecules was among cell biology's successes between 1965 and 1975. This collection of phospholipids, sterols, and proteins has been structured into a fluid crystal, an arrangement that promotes cell proliferation. The membrane's viscosity has been measured to be 100 times that of water, making it analogous to thinner oil (Beney & Gervais, 2001). On the surface of the bilayer, the phospholipid molecules disperse easily. Several membrane proteins have this freedom of mobility as well, although certain are locked in the membrane due to interactions with the cytoskeleton of the cell. Phospholipids that have recently been created easily integrate into the current cell membrane. Extrinsic proteins seen on the cell membrane's inner surface were produced on free, or unattached, ribosomes, which were then released into the cytoplasm and brought to the membrane. Intrinsic proteins have been implanted throughout their biosynthesis on ribosomes constrained to the ER (Harada, 2005).

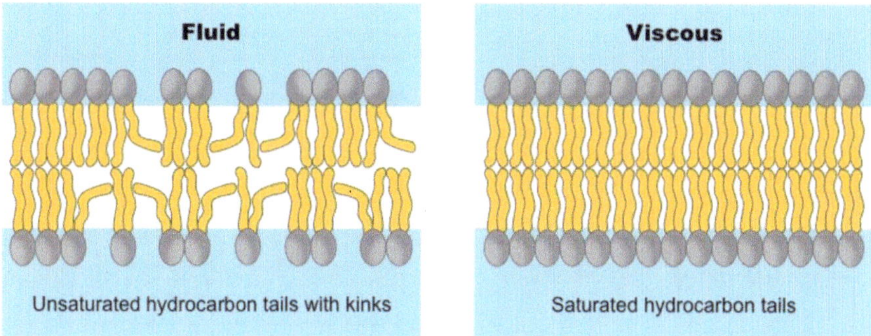


**Figure 2.7.** Because shorter fatty acid tails are less viscous and much more responsive to changes in kinetic energy, they would enhance fluidity. Double-bond lipid chains (unsaturated fatty acids) feature kinked hydrocarbon tails which make packing them together more difficult.

**Source:** https://ib.bioninja.com.au/standard-level/topic-1-cell-biology/13-membrane-structure/membrane-fluidity.html

## 2.2.5. Transport Across the Membrane

The cell membrane's chemical makeup makes it very elastic, making it an ideal border for rapidly growing and duplicating cells. The membrane works as a strong barrier on both sides, allowing certain dissolved compounds, or solutes, to get through while preventing others from doing so. While certain tiny molecules and lipid-soluble molecules may get through the membrane, the lipid bilayer efficiently repels the big, electrically charged ions and water-soluble molecules that the cell needs to survive (Postnov & Orlov, 1985). Specific classes of intrinsic proteins transport vital substances via a variety of transport systems: some are open channels, which also allow ions to enter the cell directly; others are "facilitators," which help solutes diffuse past the lipid screen via a little-understood chemical transformation; and still others are "pumps," which force solutes through the membrane even when they aren't concentrated enough to diffuse instantly. When a membrane closes and opens, particles that are too large to be dispersed or pumped are commonly consumed or disgorged intact (Rapoport et al., 1996).

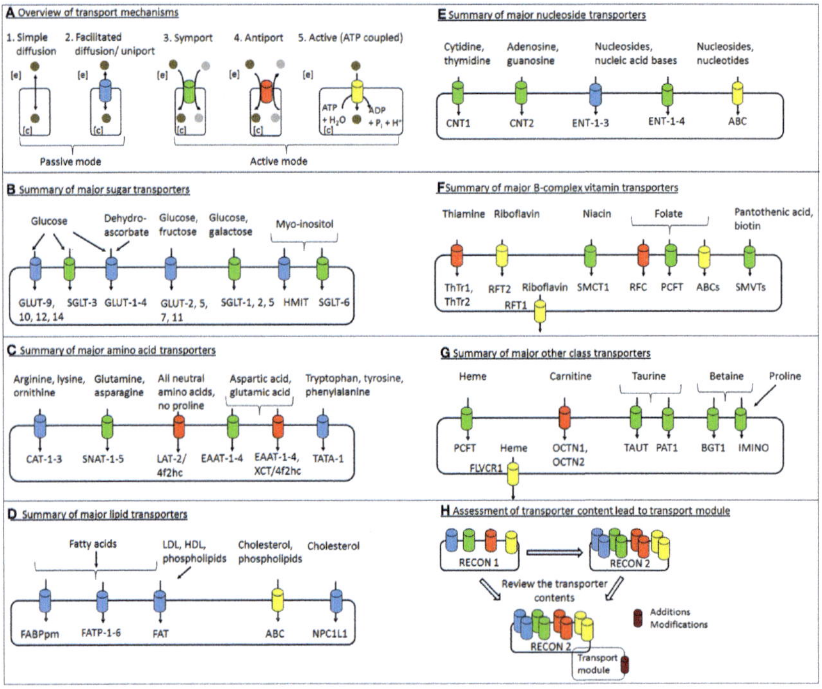


**Figure 2.8.** Transport occurs across the cell's membrane.

**Source:** Sahoo S, Aurich MK, Jonsson JJ and Thiele I (2014) Membrane transporters in a human genome-scale metabolic knowledgebase and their implications for disease. Front. Physiol. **5**:91. doi: 10.3389/fphys.2014.00091

Diffusion has been the concept that governs the flow of solutes throughout the cell membrane. A soluble material diffuses downward the gradient of concentration as per this concept; such that, assuming no energy from an external supply, it goes from a higher concentration point to a lower concentration point. Diffusion proceeds along the constantly diminishing gradient till equilibrium is attained, where at the time there has been the same concentration in both directions and random, identical diffusion on both sides (Devuyst & Rippe, 2014).

A solute with a higher concentration has higher free energy, which means it can do more "work" than a solute with a lower concentration. The solute loses free energy while conducting diffusion work, and while it achieves stability at a low concentration, it would be not capable to go back automatically (by its energy) to its previous higher concentration (Cirillo, 1962). The solute can be restored to its prior concentration and condition of higher free energy by adding energy from an outside supply (for instance, by the operation of an ion pump). Such "coupling" of work procedures is essentially a free transfer of energy from the thrust to the solute, that may replicate the work of diffusion (Wilson & Pohorille, 1996).

Many biologically important chemicals have differing concentrations within and exterior of the cell, resulting in the gradients of concentration down which solutes freely disperse if they may pass through the two-layer lipid (Dellago & Hummer, 2006). Routes of membrane and diffusion catalyst passively convey them across the membrane; the diffusing solutes alone power the modifications in the proteins that aid diffusion. Some solutes should stay at various concentrations on either hand of the membrane for the cell to operate properly; if they achieve equilibrium via diffusion, they should be pushed back their gradients via active transport. The energy needed for transport is coupled to the electricity generated via the metabolism of cells or via the diffusion of many different solutes via membrane proteins that serve as pumps (Visser et al., 2007).

### 2.2.6. Permeation

Permeation has been the diffusion of a material in solution crossways a barrier. The rate during which physically related chemicals pass through the cell membrane via permeation varies greatly. Alcohols and water, on either side, penetrate many membranes within a sec, but polymers of sugar and proteins don't. The lipid bilayer, which provides the membrane with its extraordinary permeability, causes this fluctuation. Permeability refers to the rate at which a chemical in solution flows through a membrane (Müller-Plathe, 1994).

For all of the cell membranes that are investigated in the lab, permeability rises in lockstep with the capacity to break down organic solvents. Researchers have established that permeability is regulated by the fatty acid inner of the lipid bilayer rather than its phosphoryl exterior due to the regularity of such a parallel. Inorganic solvent partition coefficients are used to quantify the ability of a substance to dissolve in organic solvents rather than in water (Rogers, 1985). A material's partition coefficient will be higher if it is more solubilized than soluble; if a substance's partition coefficient is higher than its permeability to that specific substance, the membrane will be more permeable to that specific substance. The solubility of hydroxyl, the solubility of carboxyl, and the solubility of amino groups in water, for instance, affect their solubility in organic solvents, and hence their partition coefficients. The permeability of cell membranes towards such groups is poor. Lipid-soluble methyl residues and hydrocarbon rings, on either side, with higher partition coefficients, enter cell membranes very readily, which is important in the development of pharmacological and chemotherapeutic medications (Wilschut et al., 1995).

When 2 molecules with a similar partition coefficient have a similar molecular size or mass, the one with the larger molecular mass or size would pass the membrane relatively slowly. If the molecules are tiny enough, even those with extremely lower partition coefficients may get through the membrane (Basheer et al., 2001). Water, for instance, is not soluble in organic solvents but, due to the tiny size of its molecules, it may pass through cell membranes. Because it is not a simple fluid in which molecules flow around and through the diffusing molecule, but rather an organized matrix, a form of the fixed grate, made up of the phospholipids' fatty acid chains, into which the diffusing molecule should fit, the lipid bilayer has size selectivity (Friend, 1992).

Many chemicals don't pass via the cell membrane due to lipid bilayer penetration. Certain electrically charged ions, for instance, have been attracted to organic solvents and hence have a hard time crossing cell membranes, if at all. Special openings in the membrane, known as channels, enable certain ions and tiny molecules to pass straight through the bilayer in such circumstances (Basheer, 2001).

### 2.2.7. Membrane Channels

Cell membranes have a much larger electrical conductivity as compared to a membrane bilayer consisting solely of sterols and phospholipids, according to biophysicists who measured the electric current running through them. The proteins in the cell membrane are assumed to be responsible for the increased conductivity. On recording equipment, a current that flows over a membrane frequently appears as a sequence of bursts of varying heights. Current flows via open routes that are just holes generated through intrinsic proteins crossing the bilayer lipid, resulting in such bursts. If no route is open, no substantial current travels through the membrane; if more than a route is open, several bursts have been recorded (Cooper et al., 1985).

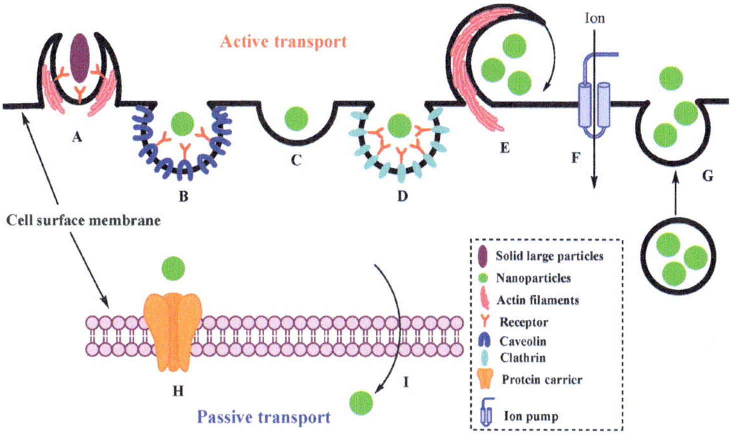


(**A**) phagocytosis, (**B**) caveolin-mediated endocytosis, (**C**) clathrin–caveolin-independent endocytosis, (**D**) clathrin-mediated endocytosis, (**E**) macro-pinocytosis, (**F**) ion pumps, (**G**) exocytosis, (**H**) facilitated diffusion, and (**I**) simple diffusion with reference to nanoparticles

**Figure 2.9.** Across facilitated diffusion, protein pathways allow big or polar molecules to flow through the partially permeable cell membrane.

**Source:** Sabourian P, Yazdani G, Ashraf SS, Frounchi M, Mashayekhan S, Kiani S, Kakkar A. Effect of Physico-Chemical Properties of Nanoparticles on Their Intracellular Uptake. *International Journal of Molecular Sciences*. 2020; 21(21):8019. https://doi.org/10.3390/ijms21218019

From a vast spectrum of cell membranes, a diverse range of pathways is identified and studied. Several amino acid patterns transverse the membrane in intrinsic proteins, generating a distinct pore or hole. Occasionally, some routes will close and open by themselves (Ren et al., 2007). A few are closed or opened through the chemical action of a signaling agent like acetylcholine, calcium, or glycine, while others have been closed or opened via variations in the electrical potential throughout the membrane. Pathways can have a restricted specification, enabling just sodium or potassium to flow through, or a wide specification, enabling any anions or cations *to pass through. Gap junctions are passageways that enable molecules to move across pairs of cells (*Maurel et al., 2008).

The various nerve-muscle, nerve-nerve, and nerve-gland connections that underpin neurobiological behavior are based on the regulation of routes with the ability to transport ions. The electric potential of the cellular membrane, which fluctuates depending upon the elements present in the cell's surroundings, is responsible for such processes. When a route in a membrane dividing 2 KCl solutions allows solely k+ ions, the k+ ions will flow along their gradient of concentration via the passageway. The $Cl^-$, which have a -ve charge, remain behind (Khalili-Araghi et al., 2009). Diffusion potential is the electric potential created by the dispersion of electric currents over a membrane. The concentration difference of the permeate ion throughout the membrane, among some other things, determines the size of this potential. Generally, the cell membrane comprises pathways with a broad range of ion particularities, every contributor to the overall membrane potential based on the concentration ratio and the permeability of the ion moving from it. Because channels are frequently gated, the membrane's potential is defined through which pathways are open; this, in turn, has been defined via signaling molecule concentrations and can change over time depending upon the membrane potential (Khalili-Araghi et al., 2009).

Many cells get a ten times larger sodium ion concentration exterior than inside, as well as a potassium ion concentration ratio that is the opposite. Outside of the cell, free calcium ions may be ten thousand times more concentrated than within. Consequently, potassium-, sodium-, and calcium-selective membrane pathways typically function as transmitters of nerve impulses by permitting the passage of such ions across the cell membrane and producing changes in the electric potential of the membrane. Ion diffusion poses a hazard to the ion concentrations required for cell activity. The activity of ion pumps restores the normal distribution of ions (Pohorille et al., 2005).

## 2.2.8. Endocytosis

At the beginning of the procedure, the cell membrane engulfs sections of the exterior media and creates an almost completely circular shell around the cell. The endosome is then drawn into the cell by the membrane-bound vesicle that has been drawn into the cell (Silverstein et al., 1977). There are many kinds of endocytosis: pinocytosis, in which the vesicles are smaller and comprise of fluid; phagocytosis, in which the vesicles are larger and comprise of solid materials; and receptor-mediated endocytosis, in which the material binds specifically to a receptor on the cell membrane's exterior face, triggering the procedure with which it has been engulfed. Cholesterol enters cells via the last accessible pathway (Sahay et al., 2010).

## 2.2.9. Exocytosis

Exocytosis is the process by which material generated within the cell and bundled into membrane-bound vesicles has been ejected from the cell when the vesicles fuse with the exterior cell membrane. Cell-specific protein components, neurotransmitters, and a range of other compounds are among the items being ejected (Creutz, 1992).

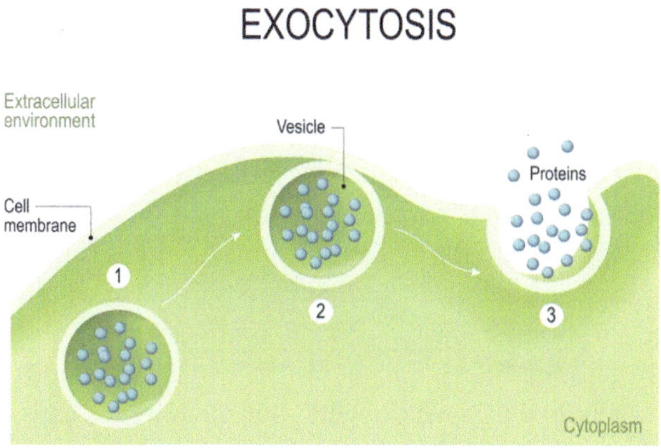


**Figure 2.10.** The process by which vesicles are transported to the cell membrane, fuse with the membrane, and their contents are released into the extracellular environment is called exocytosis.

**Source:** https://www.thoughtco.com/what-is-exocytosis-4114427

## 2.2.10. Internal Membranes

Inner membranes separate eukaryotic (cells having a nucleus) from prokaryotic cells (cells without a nucleus). Prokaryotic cells have been tiny (1 to 5 micrometers in length) and have just one cell membrane; metabolic processes are frequently restricted to sections of the membrane instead of portions of the cell's body. In comparison, most eukaryotic cells are significantly bigger, with the cell membrane accounting for just 10% less than the entire cellular membrane. Organelles, sections isolated from the cytoplasm, or cell body, via internal membranes, perform metabolic duties in such cells (Kotnik & Miklavčič, 2006).

Inner membranes are functional and structural elements of eukaryotic cells' organelles and vesicles. The mitochondrion, nucleus, and chloroplast (in plants) are the most important organelles. Peroxisomes, glyoxysomes, peroxisomes, and lysosomes are the residual organelles that contain highly reactive by-products and enzymes (Novak et al., 2009). Inner membranes create the maze-like endoplasmic reticulum, which is responsible for the synthesis of cell membrane lipids and proteins, as well as the Golgi apparatus, which is responsible for the modification and transport of proteins, lipids, and carbohydrates. Inner cell membranes may also create transport and storage vesicles, as well as plant cell vacuoles. Every membrane structure has a distinctive content of lipid and protein that allows it to perform certain activities (Uittenbogaard et al., 2002).

## 2.2.11. General Properties and Functions of Membrane

The membranes of certain organelles, including the cell membrane, include transport proteins called permeases which facilitate the communication of chemicals across organelles. Permeases inside the lysosomal membrane, for instance, enable amino acids produced within the lysosome to enter into the cytoplasm and be utilized for protein synthesis (Pucadyil et al., 2009). The membrane budding mechanisms of exocytosis and endocytosis, which have been fundamentally similar to the cell membrane, have been used to communicate among organelles.

Degradative and biosynthetic activities in various organelles, on either side, can necessitate circumstances that are vastly various from those in other organelles or the cytoplasm. Inner membranes keep such distinct situations apart by separating them. For example, the pH of the interior of lysosomes is higher than that of the cytosol (pH 5 vs. pH 7) and is controlled by particular proton-pumping transfer proteins in the membrane of the lysosome (Radons & Multhoff, 2005).

Organelles also help to prevent enzyme operations that compete with one another from interfering with one another. Important proteins, for example, have been produced in the RER and the cytosol, whereas unnecessary proteins are degraded in the lysosomes and, to a lesser extent, in the cytosol, according to the findings (Chen & Li, 2013). Additionally, fatty acids are synthesized in the cytosol and subsequently degraded in the mitochondria to make ATP or destroyed in the peroxisomes to generate heat, which is then recycled back into the cytosol. Such processes should be kept distinct from one another and should not be used together. It is also possible that organelle membranes, in addition to sequestering potentially hazardous by-products or enzymes in their respective membrane-bounded chambers, avoid potentially fatal by-products or enzymes from attacking responsive components in other sections of the cell (Cronan Jr & Gelmann, 1975).

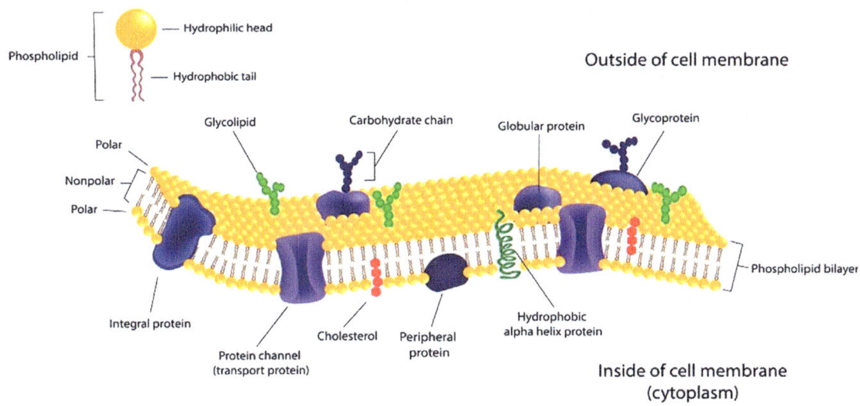


**Figure 2.11.** This diagram depicts the function and structure of a cell membrane.

**Source:** Jiménez-Jiménez C, Manzano M, Vallet-Regí M. Nanoparticles Coated with Cell Membranes for Biomedical Applications. Biology. 2020; 9(11):406. https://doi.org/10.3390/biology9110406

Eukaryotic cell interior membranes are chemically and physically distinct from the external cell membrane. They, such as the external membrane, are made up of a phospholipid bilayer in which some membrane proteins have been bound or embedded. Cholesterol, phospholipids, and glycolipids, the 3 main lipids that make up the exterior membrane, are also found in the interior membranes, although in varying quantities (Hanany et al., 2008). Phospholipids are the fundamental lipids that make up all cell membranes. Cholesterol that assists in the stability and fluidity of all membranes has been found at roughly 25% of the quantity in the external membrane in interior membranes. Glycolipids are just minimal elements of interior membranes, but they account for around 5% of the lipid in the outer membrane (Aronson, 1985).

## 2.3. CELLULAR ORGANELLES AND THEIR MEMBRANES

### 2.3.1. Vacuole

Vacuoles are membrane-bound vesicles found in the majority of plant cells. The cell sap, carbohydrates and a water solution of salts that is kept at a higher concentration by active ion transport via permeases in the vacuole membrane are kept in the vacuole. Proton pumps are also in charge of keeping protons in the vacuole interior at a high level (Glaumann et al., 1981). Water may enter the vacuole by osmosis at such high quantities, causing the vacuole to grow and turgor, the cell membrane is forced against the cell wall as a result of hydrostatic pressure, causing the cell to burst. Turgor causes rigidity in live plant tissue.

In a mature plant cell, a single vacuole may occupy up to 90% of the cell volume; immature cells generally include numerous smaller vacuoles (Håkansson et al., 2001).

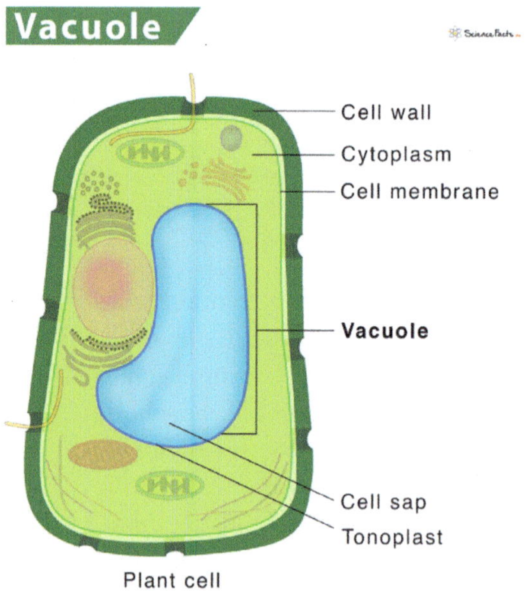


**Figure 2.12.** Vacuole Structure.

**Source:** https://www.sciencefacts.net/vacuole.html

## 2.3.2. The Lysosome

To safeguard the other elements of the cell from arbitrary disintegration, potentially harmful hydrolytic enzymes that act in acidic circumstances (pH 5) have been classified in the lysosomes. A solitary phospholipid bilayer membrane binds lysosomes (Molino et al., 2017). They come in a variety of sizes and have been generated by the fusion of the Golgi-derived vesicles having cell surface-derived endosomes. Hydrolases that destroy nucleic acids, proteins, glycolipids, lipids, and glycoproteins are found to exist in lysosomes. Hydrolases have been particularly active in the lysosomes' acidic environment. Permeases transfer lipids and amino acids all over the lysosomal membrane used in biosynthesis once the material has been decomposed. A leftover body is the leftover detritus that usually stays inside the lysosome (Dell'Angelica et al., 2000).

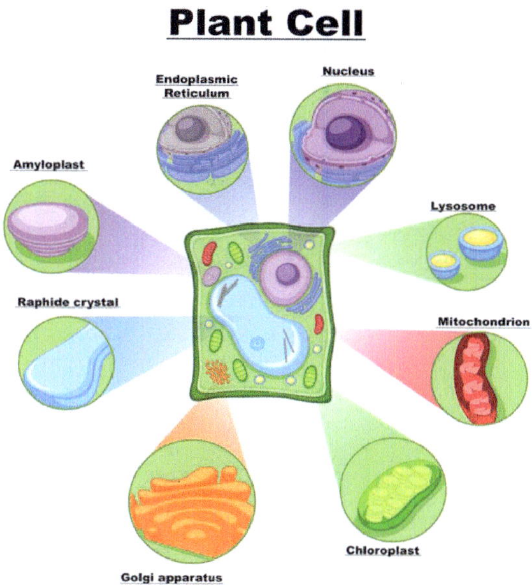


**Figure 2.13.** Lysosomes are membrane-bound organelles found in both plant and animal cells' cytoplasm.

**Source:** https://biologywise.com/lysosome-structure

## 2.3.3. Microbodies

Microbodies are half to one micrometer in diameter, roughly spherical, and bound via a single membrane. There have been several kinds, with the peroxisome being the most common. Peroxisomes acquire their name from $H_2O_2$, a reactive step in the chemical breakdown procedure of microorganisms. Peroxisomes include type II oxidases, which have been enzymes that utilize molecular oxygen to oxidize organic compounds (Vigil, 1970). Hydrogen peroxide has been produced as a byproduct of such reactions, which is then metabolized into oxygen and water through the enzyme catalase that is found in abundance in peroxisomes. Peroxisomes also comprise enzyme systems that deteriorate a variety of lipids.

The plant glyoxysome is peroxisome which also includes glyoxylate cycle enzymes that are essential for fat to carbohydrate conversion (Gietl, 1992).

## 2.4. THE NUCLEUS

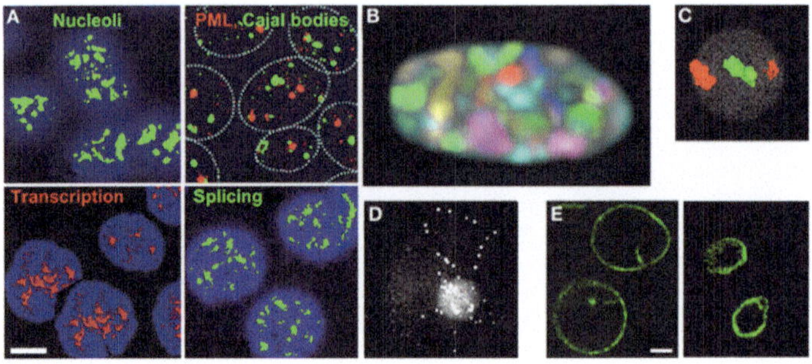


**Figure 2.14.** Nucleus Structure.

(A) Immunofluorescence (reproduced from [20]) of compartments. (B) Discrete territories of the 24 chromosomes in a human fibroblast nucleus labeled with different fluorochrome combinations (reproduced from [21]). (C) Gene-poor chromosomes ([18], red) are located more peripherally than gene-rich chromosomes ([19], green) in the nucleus of human lymphocytes (reproduced from [22] by permission from Macmillan Publishers Ltd, 2001). (D) A loop of DNA containing ~220 kb of the human dystrophin gene visualized by fluorescence in situ hybridization (FISH) (reproduced from [23] by permission of Oxford University Press). (E) The nuclear lamina confines and compresses the contents of the nucleus.

**Source:** Hancock R (2014) Structures and functions in the crowded nucleus: new biophysical insights. Front. Phys. **2**:53. doi: 10.3389/fphy.2014.00053

In all eukaryotic species, the nucleus serves as the cell's information hub, and it is surrounded by a nuclear membrane to protect it from outside influences. While the nuclear envelope protects the nucleus from the surrounding cytoplasm, it also contains the double-stranded, spiral-shaped DNA molecules that hold the genetic data that allows the cell to preserve its exceptional character as it divides and develops (Shibata et al., 2009).

The existence of a nucleus differentiates multi-cellular eukaryotic cells from single-celled prokaryotic organisms like bacteria. Because prokaryotes lack nuclei, their DNA has been kept in a similar chamber as the rest of its cellular elements, unlike higher organisms (Somosy, 2000).

The nucleus' basic task is to articulate specific subsets of the genetic data programmed in the two fold helix of deoxyribonucleic acid. RNA and DNA are both types of nucleic acids, and genes are segments of deoxyribonucleic acid that code for the creation of a certain protein from an amino acid sequence. Although, the deoxyribonucleic acid information has not been directly translated into proteins (Xiong et al., 2006). A sequence of mRNA molecules is produced, each of which encodes information for a specific protein. This is the initial step in the transcription process. The mRNA molecules are subsequently transported into the cytoplasm by way of the nuclear envelope, where they are translated and employed as templates for the biosynthesis of certain proteins, depending on the situation (Field et al., 2015).

Nuclei are responsible for a variety of functions, including not only the synthesis of messenger RNA (mRNA) for several hundreds of proteins but for controlling the quantities of protein that are produced and transported to the cytoplasm. More importantly, in every cell type, the amounts of every kind of mRNA that is delivered to the cytoplasm should be controlled distinctly. More importantly, the mRNA, the nucleus generates and ejects a wide variety of other RNA types, each of which plays a role in the many pathways of protein production (Frey et al., 2006).

## 2.4.1. Structural Organization of the Nucleus

### *2.4.1.1. DNA Packaging*

However, even though the average human cell's nucleus is just six μm in diameter ($6 \times 10^{-6}$ meters), it carries approximately 1.8 meters of DNA. It is spread throughout 23 pairs of chromosomes, which are composed of a single DNA molecule measuring around forty millimeters (1.5 inches) in length. A scale model that has been magnified a million times may be used to illustrate the huge packaging challenge this presents (Ward & Coffey, 1991). This scale depicts DNA molecules as thin strings two millimeters thick, and an average chromosome will comprise forty kilometers (twenty-five miles) of DNA on average. The nucleus will hold 1,800 kilometers of DNA despite having a diameter of just six meters.

Such components should be structured in such a way that they may be replicated into RNA with accuracy and selectivity when they are transcribed into RNA. Deoxyribonucleic acid is not merely packed or twisted into the nucleus such as a ball of thread; rather, it is woven into the nucleus (Steven

Ward & Coffey, 1991). However, it is structured into a finely packaged form through molecular interaction with certain nuclear proteins. The interaction of DNA and proteins results in the formation of chromatin, which is a dense, compressed fiber. An illustrative case of the planned folding and compaction that chromatin may go through is shown during the division of cells when the chromatin of every chromosome condenses to form a single chromosome and then divides into 2 daughter cells (Balhorn et al., 1993).

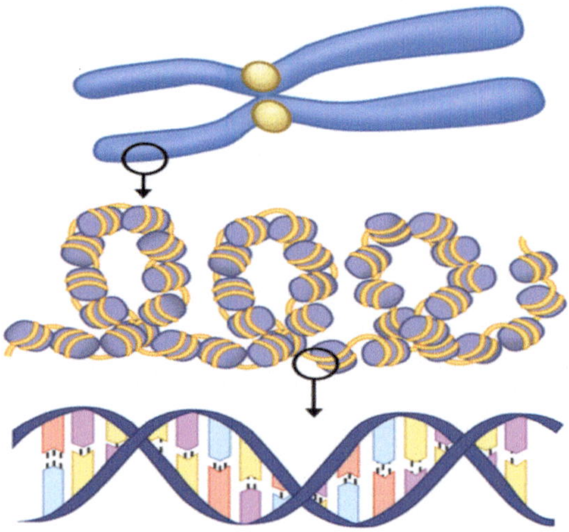

**Figure 2.15.** DNA packaging. Chromosome (top), chromatin coiled around histones forming nucleosomes (middle), DNA double helix (bottom).

**Source:** https://theory.labster.com/dna-packaging/

### *2.4.1.2. Nucleosomes*

DNA is compacted via winding it around a sequence of tiny proteins known as histones, which serve to hold the DNA together. Histones are made up of +ve charged amino acids which bind securely to DNA and neutralize the -ve charges on the strand of the DNA molecule. Histones are classified into five types. An octamer is an 8-part core around which 2 turns of deoxyribonucleic acid have been wrapped (Bock et al., 1994). Four of these, designated H2A, H2B, H3, and H4, contribute 2 molecules each to make an octamer. The nucleosome is the structure formed as a result of this reaction. The deoxyribonucleic acid enters and exits a succession of nucleosomes, tying them together as well as beads on a string in lengths that range among various species of the organism or even among various kinds of cells inside a species of the organism. Nucleosomes are wound together into a solenoid structure by the 5th histone, known as H1, which is found in the nucleus. It is thought that each nucleosome is bound by one H1 molecule and that a chain of H1 molecules is responsible for coiling a string of nucleosomes into the solenoid makeup that is found in DNA fibers (chromosomes) (Li & Zhu, 2015).

Nucleosomes not only have the effect of neutralizing the charges on DNA, but they also have additional effects. For starters, they are a cost-effective method of packing. When DNA is twisted into nucleosomes, it is compressed by 6 times, and when the nucleosomes have been curved into a solenoid chromatin fiber, it is squeezed by 40 times. Furthermore, the wrapping of DNA into nucleosomes enables inactive DNA to be folded away into unreachable conformations, increasing the selectivity with which genes are expressed (Sanders, 1978).

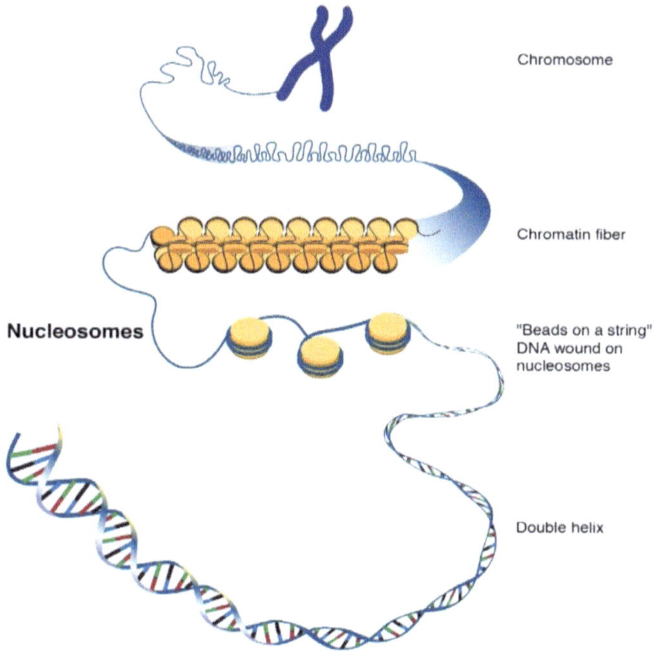


**Figure 2.16.** Nucleosome Structure and Function.

Source: https://www.genome.gov/genetics-glossary/Nucleosome

### 2.4.1.3. Chromatin Fiber Organization

According to certain studies, chromatin is arranged into a sequence of enormous radial loops which are anchored to certain scaffold proteins. Every loop is made up of a nucleosome chain and could be linked to genomic organizing units. The chromatin loops are arranged in a radial pattern that compacts DNA by a factor of a thousand. The looped chromatin fiber is further compacted by coiling it into an opaque structure known as a chromatid, of which 2 constitute the chromosome. Such coiling causes ten thousand times DNA compaction upon cell division (Dillon, 2006).

### 2.4.1.4. The Nuclear Envelope

The nuclear envelope is made up of two phospholipid bilayers, one on the outside and one on the inside. The rough endoplasmic reticulum (RER) lumen is connected to the tiny space between the 2 layers, and the exterior layer is an extension of the RER's external face (Gerace & Blobel, 1982).

The nuclear lamina, a protein lining that covers the inside surface of the nuclear envelope, is responsible for the binding of chromatin and other nucleoplasmic substances to the nuclear envelope. Several nuclear pores perforate the whole nuclear envelope. Larger molecules may pass through such transport routes without being affected, even though they are entirely permeable to small molecules up to the size of the tiniest proteins (Monneron & Bernhard, 1969). Every pore is bounded by the nuclear pore complex, a complex protein structure that selects which molecules are allowed to enter the nucleus via each pore. The nucleotide basic elements of RNA and DNA, and also adenosine triphosphate, which gives energy for genetic material synthesis, enter the nucleus through the pores. Large proteins, such as histones, should also pass through the pores. On the surface of such molecules is a unique pattern of amino acid which signifies nuclear pore complex admission. The complexes also control the outflow of ribosome subunits and RNA from the nucleus (Webster et al., 2009).

In prokaryotes, DNA is arranged in loops and coupled to tiny proteins that resemble histones, however, such structures are not surrounded by a nuclear membrane (Burke & Stewart, 2002).

Telomeres interact with the envelope by RAP1 and proteins in the inner nuclear membrane (INM) like SUN1. These interactions may be stabilized by other connections across the lumen to outer nuclear membrane (ONM) nesprins that connect to cytoplasmic filaments. The intermediate filament lamin polymer that lines the INM interacts with core histones to also tether chromatin to the NE.

Many specific interactions though the chromatin partners are yet to be identified (?). FHeterochromatin proteins such as LBR binding heterochromatin protein 1 (HP1) are also involved.

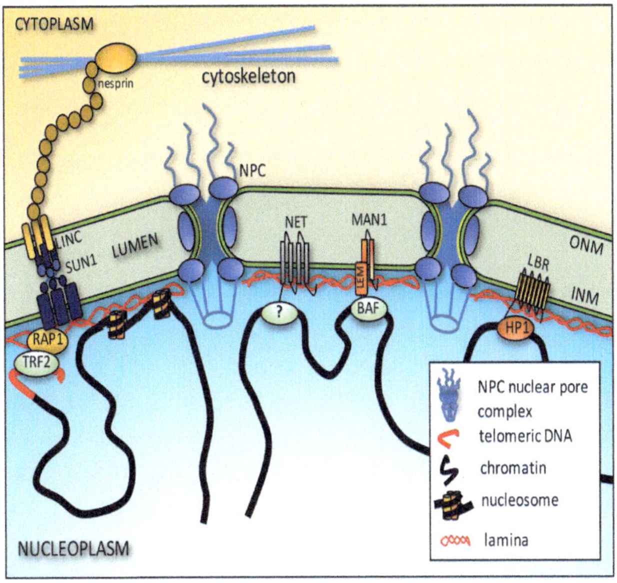


**Figure 2.17.** The Nuclear Envelope is depicted graphically.

**Source:** Czapiewski R, Robson MI and Schirmer EC (2016) Anchoring a Leviathan: How the Nuclear Membrane Tethers the Genome. Front. Genet. *7:82*. doi: 10.3389/fgene.2016.00082.

## 2.4.2. Genetic Organization of the Nucleus

### *2.4.2.1. The Structure of DNA*

The genetic structure among most organisms shares certain characteristics. The double-stranded DNA is the first. This molecule's strands are made up of nucleotides, which are made up of a sugar-phosphate complex coupled to one of 4 nitrogen-consisting bases. The backbone of the strand is made up of sugar-phosphate molecules that are linked together. Every base strung along the backbone of the DNA molecule has been chemically linked to a comparable base on the parallel strand. The chemical interaction of the base pairs bends the two strands into a helical or spiral shape, similar to how rungs join the 2 aspects of a ladder (Watson & Crick, 1953).

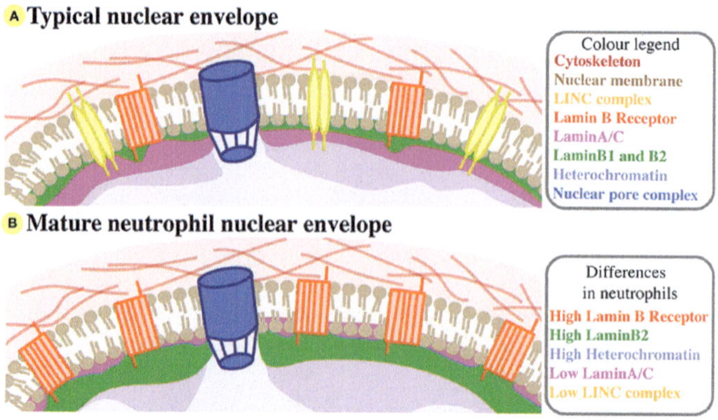


A) A typical nuclear envelope comprises of the nuclear membrane bilipid layer (brown) with membrane proteins like the LINC complex (yellow) and Lamin B Receptor (orange), and nuclear pore complexes (blue). The nuclear envelope interacts with the cytoskeleton (red). The nuclear lamina is a structual mesh formed of LaminA/C (pink) and B-type lamins (green) (B) The nuclear envelope of mature neutrophils has very low levels of LaminA/C and LINC, but increased Lamin B receptor, peripheral heterochromatin, and LaminB2.

**Figure 2.18.** Nucleus function and structure.

**Source:** Manley HR, Keightley MC and Lieschke GJ (2018) The Neutrophil Nucleus: An Important Influence on Neutrophil Migration and Function. Front. Immunol. *9:2867.* doi: 10.3389/fimmu.2018.02867

Cytosine, adenine, thymine, and guanine are the 4 nucleotide bases. Deoxyribonucleic acid is formed by thousands of such bases strung together in an infinite number of ways. The genetic data is included in the pattern of bases, with every pattern specifying the amino acids pattern to be joined into proteins. A gene is a nucleotide pattern that can encode only one protein. Certain sequences which don't encode proteins are interspersed having genes along the DNA molecule. Certain untranslated areas, for instance, govern the activity of nearby genes by indicating when enzymes start and stop transcribing DNA into RNA (Lindahl, 1993).

### *2.4.2.2. Rearrangement and Modification of DNA*

When nucleotide pattern in DNA is rearranged or modified, they have been considered exceptions to the genetic expression laws, and they may cause major structural variations or cell functions at times. Various genes are responsible for the specific functions and structures of the various types of cells in the body. This doesn't imply that the genetic data set contained within the body cells differs from one another (Bétermier & Duharcourt, 2014). However, the whole DNA composition of the chromosomes is typically copied perfectly from generation to generation for every cell, and the genetic composition and organization of the chromosomes are remarkably comparable across various kinds of cells within the same organism. The differentiating of cells may occur as a consequence of this lacking the loss or irreversible deactivation of superfluous genes, a discovery that is supported by the existence of particular genes in a variety of mature tissues. Examples include the fact that normal duplicates of the genes for haemoglobin have been found in identical amounts in red blood cells that produce haemoglobin, as they are in a variety of certain kinds of cells, that don't produce haemoglobin (Bernt & Armstrong, 2011).

Just a few unambiguous instances of planned, reversible DNA modifications in the DNA of developing tissues have been discovered in certain species despite the overall uniformity and consistency of genetic composition across whole cells of the organism, according to recent research. The immune systems of animals are home to one of the most spectacular DNA rearrangements known to science (Levin-Klein & Bergman, 2014). The production of a wide range of antibodies via lymphocytes, the body's defense versus invasion via foreign organisms, is essential in preventing invasion. Proteins called antibodies adhere to invading molecules or organisms and either deactivate or indicate the death of the invaders. In the lymphocyte nucleus, the binding sites on every antibody molecule have been produced via two amino acid chains, one heavy and one light that is encoded via separate portions of the lymphocyte nucleus' DNA. Such DNA sections go through a substantial deal of rearrangement, which results in the creation of a large number of different antibodies (Fang et al., 2012). When some invading species, like the trypanosome parasites that cause sleeping sickness, infect their hosts' cells, their DNA is changed to evade the flexibility of their hosts' antibody manufacturing to the maximum degree feasible. Glycoprotein coats the parasites, providing them with protection from the surrounding environment (a protein having sugars attached). However, while host organisms may be able to overcome infection over time by making antibodies against the parasites' glycoprotein coat, this reaction has been predicted and evaded by a specific reorganization of the trypanosomes' deoxyribonucleic acid programming the glycoprotein, which results in the surface introduced to the host's immunity system changing constantly (Pérez et al., 1997). The modifications of epigenetic, heritable alterations which occur on the sugar-phosphate side of bases in DNA and therefore don't create rearrangements in the pattern of the DNA, are also discovered through careful analyses of gene structure. An instance of epigenetic alteration is the insertion of a methyl group into cytosine bases, which is a type of DNA modification. It is thought to occur in the deactivation of genes that are not required to be expressed in a specific type of cell. When cytosine is methylated, one of the most essential characteristics of the process is its capacity to be replicated, which means that methyl groups in the DNA of a dividing cell would outcome in methyl groups in identical positions in the DNA of both daughter cells (Buckhout-White et al., 2018).

### *2.4.2.3. Genetic Expression by RNA*

The genetic information transcription from DNA to RNA, followed by the translation of that code from RNA to a protein, has the greatest impact on the control of genetic data. The gene expression procedure occurs in phases, with the possibility for an additional type of cell differentiation at every stage (Iwamoto et al., 2009).

In the DNA molecule, as previously mentioned, genetic data is programmed in the patterns of the four nucleotide bases that make up the molecule. With the addition of uracil to DNA, the thymine nucleotide is altered, and one of the 2 DNA strands is accurately translated into mRNA. In its linking sugar-phosphate chain, RNA has a bit more distinct sugar element (ribose) than DNA (deoxyribose). Whereas DNA, remains stable across a whole cell's existence and may even be transmitted down via generations of several cells, RNA is unbalanced. It is constantly decomposed and replenished, allowing the cell to alter its protein production patterns (Culbertson, 1999).

Aside from mRNA, which encodes proteins, the nucleus produces additional types of RNA. The rRNA is a component of ribosomes that is transferred to the cytoplasm to aid in the translation of mRNA data into proteins. The rRNA is produced in the nucleolus, a specialized portion of the nucleus that appears like a densely populated area inside the nucleus and includes the genes that code for ribosomal RNA (Kumar et al., 2013). Ribosome subunits are formed via the assembly of rRNA and ribosomal proteins, which takes place at this location as well. Ribosomal proteins are synthesized in the cytoplasm and then transported to the nucleus, where they are sub-assembled in the nucleolus before being released. Following that, the subunits are returned to the cytoplasm, where they will be assembled again. Transfer RNA (tRNA) is a kind of RNA that is synthesized in the nucleus and serves as an adaptor, connecting individual amino acids to the nucleotide triplets of messenger RNA throughout the protein production cycle (Battle et al., 2014).

### 2.4.2.4. RNA Synthesis

RNA polymerase enzymes perform RNA synthesis. There have been 3 kinds of RNA polymerases in higher organisms: I, II, and III (or sometimes A, B, and C). Each one is a multi-subunit complicated protein. RNA polymerase 'I' is active in the nucleolus, in which the genes that encode such rRNA molecules lie since it synthesizes 3 of the 4 forms of rRNA (named 18S, 28S, and 5.8S RNA). RNA polymerase II produces mRNA, however, its early products are bigger precursors known as heterogeneous nuclear RNA, that are finished later (shown below the Processing of mRNA). RNA polymerase III produces tRNA and 5S RNA, which is the 4th RNA element of the ribosome (Krieg & Melton, 1987).

All 3 polymerases begin RNA synthesis at specified places on DNA and progress down the molecule, progressively connecting chosen nucleotides till they reach the gene's end and stop the expanding RNA chain. The higher-energy phosphate links found in the nucleotide precursors of RNA provide energy for the synthesis of RNA. The ultimate RNA product is largely made up of sugar, base, and one phosphate, however, the basic elements are made up of base, sugar, and 3 phosphates (Sola et al., 2015). For every nucleotide integrated into RNA, 2 phosphates are cleaved and eliminated throughout the synthesis. The phosphate bonds emit energy, which is utilized to connect the nucleotides. The nucleotide sequence combined into a developing RNA chain is determined by the pattern of nucleotides in the DNA template: every adenine in DNA indicates uracil in RNA, every cytosine requires guanine, every guanine requires cytosine, and every thymine defines adenine in DNA. The data contained in every gene is transcribed into RNA, which is then translated by the cytoplasm's protein-synthesizing machinery (Kao et al., 2001).

The nucleotide DNA pattern provides additional information dictating the order of amino acids to be polymerized into proteins. Short nucleotide sequences, for instance, dictate the starting site of every RNA polymerase, indicating when and where the synthesis of RNA should take place. The starting site sequences for RNA polymerases I and II are located slightly ahead of the genes (Romano et al., 2020). RNA polymerase III, on either side, has the corresponding information inside the gene, inside the DNA area to be replicated into RNA. A promoter is a DNA region that serves as a starting site. Several genes' promoters share certain nucleotide sequences, while they contradict the others. Distinct proteins known as transcription parameters identify the changes in pattern and are required for the production of specific kinds of genes. Differences in gene expression across various kinds of cells are due to transcription factor specificity (Sosunov et al., 2003).

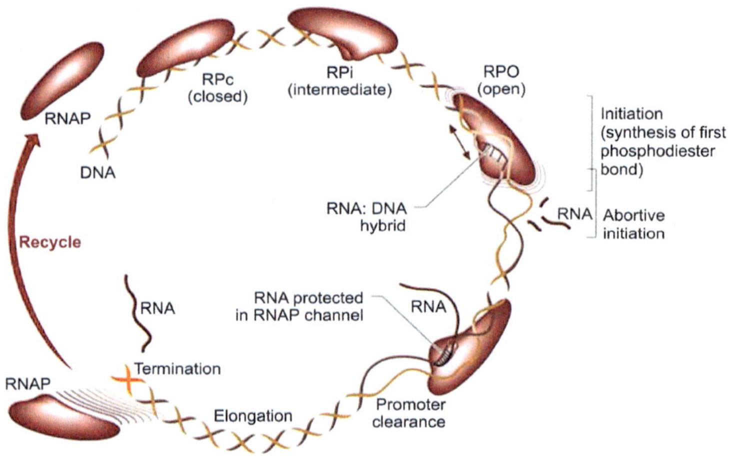


**Figure 2.19.** RNA Synthesis cycle.

**Source:** https://www.researchgate.net/figure/fig3_227979950

### *2.4.2.5. Processing of mRNA*

Before mature molecules have been discharged from the nucleus, mRNA precursors go through a complicated sequence of modifications after and during synthesis. A customized nucleotide is 1st added to the beginning of the RNA molecule in a process known as capping. Subsequently in the cytoplasm, this cap connects to a ribosome. The chemical breakage of the RNA chain, not the separation of the RNA polymerase from DNA, stops the production of mRNA. A simple polymer of adenosine residues is inserted into the cleaved terminals of several (though not all) forms of mRNA (Bentley, 1999).

Astonishing findings in 1977 indicated that sections of newly generated RNA molecules are cut out and destroyed, in addition to such termini alterations. In several genes, introns are nucleotide sequences that are interspersed between the protein-coding regions. Before they may be discharged from the nucleus like a functioning mRNA, such introns should be removed from the RNA copy. Inside a gene, the size and number of introns vary widely, ranging from none to more than fifty. The lengths of such intervening patterns can be greater than the lengths of the protein-coding regions (Singh et al., 2009).

RNA splicing, or the elimination of introns, is believed to be mediated via small nuclear ribonucleoprotein particles (snRNPs). The RNA sequences of these particles are comparable to the junctions among introns and neighboring coding areas. An snRNP bends the intron into a loop by binding to the junction ends. The loop is then excised, and the coding sections are spliced (Müller-McNicoll et al., 2016).

### *2.4.2.6. Regulation of Gene Expression*

Even though all of an organism's cell nuclei have the same genes, there are major differences between the specialized cell types found in the body. The mechanisms that control the quantities and activities of mRNA both after and during their production in the nucleus, in particular, may be responsible for these variations; in particular, processes that control the quantities and activities of mRNA both after and during their biosynthesis in the nucleus may be responsible for these variations (O'Brien & Granner, 1996).

### *2.4.2.7. Regulation of RNA synthesis*

Changes in the structure of chromatin mediate the 1st level of regulation. A gene should be organized into a structurally different kind of active chromatin to be transcribed. The 2nd level of regulation is performed by altering the frequency during which an RNA polymerase transcribes a gene in the active state into RNA. There has been evidence for the synthesis of RNA regulation at such levels, such as in response to hormone stimulation. Protein factors are thought to regulate at both levels, for instance, via binding to specific promoter DNA sequences surrounding the transcribed gene (Allfrey et al., 1964).

### *2.4.2.8. Regulation of RNA after Synthesis*

Following biosynthesis, RNA molecules are subjected to careful processing, which concludes in a subset of RNA molecules being exported to the cytoplasm. Moreover, the stability of a specific form of mRNA in the cytoplasm may be controlled (Wolter et al., 1997). The hormone prolactin, for instance, induces a twofold rise in the amount of the synthesis of mRNA in tissue, but also causes a 17-time rise in mRNA lifespan, therefore the major source of enhanced synthesis of protein in this situation is the longer accessibility of mRNA. On the other hand, proof exists for the selective instability of certain mRNA, like histone mRNA, that is quickly degraded while DNA replication is disrupted. Ultimately, there have been several instances of selective control of mRNA to protein translation (Crick & Hughes, 1950).

## 2.5. CYTOPLASM

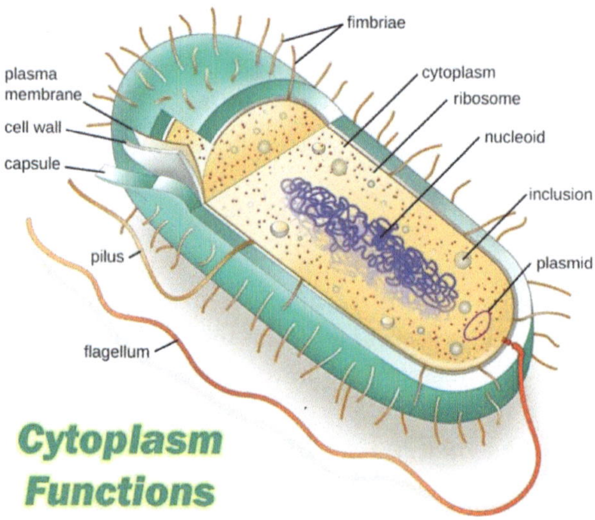


**Figure 2.20.** Cytoplasm Function and Structure.

**Source:** https://www.bioexplorer.net/cytoplasm-functions.html/

The cytoplasm is the portion of the cell that lies outside of the nucleus and fills the space between the plasma membrane and the nuclear envelope. It consists of the following components (Görlich & Kutay, 1999):

- The cytosol is the cytoplasm's largest fluid element.
- Organelles ("small organs") are metabolically active structures that can be membranes (like mitochondria) or non-membranous protein complexes that float about (like proteasomes and ribosomes).
- Eukaryotic cells' form and movement are determined by the cytoskeleton, a protein element of the cytoplasm.

- Inclusions are cytoplasmic entities that are not normally encased by a membrane. They are made up of a variety of elements including pigment granules, crystals, glycogen, lipids, and other solid wastes that have been preserved.

## 2.5.1. Cytosol

The cytoplasmic matrix is an aqueous gel that surrounds the cytosol. Inorganic ions (sodium ion, potassium ion, and calcium ion), as well as organic compounds like carbohydrates, intermediate metabolites, proteins, lipids, and RNAs, make up the matrix (ribonucleic acids) (Montanaro et al., 2008). The cytoplasmic compartment's rate of metabolic activity is influenced by the concentration of solutes in the matrix, which is controlled by the cell. Hundreds of enzymes, as well as all of the equipment that converges on the ribosomes for the synthesis of protein, electrolytic ions, oxygen, low-molecular-weight substrates, carbon dioxide, waste products, and metabolites, are all found in the cytosol. All diffuse freely or are attached to proteins through the cytosol, entering or exiting organelles in which they are needed or generated (Cross et al., 2009).

## 2.5.2. Cytoplasmic Organelles

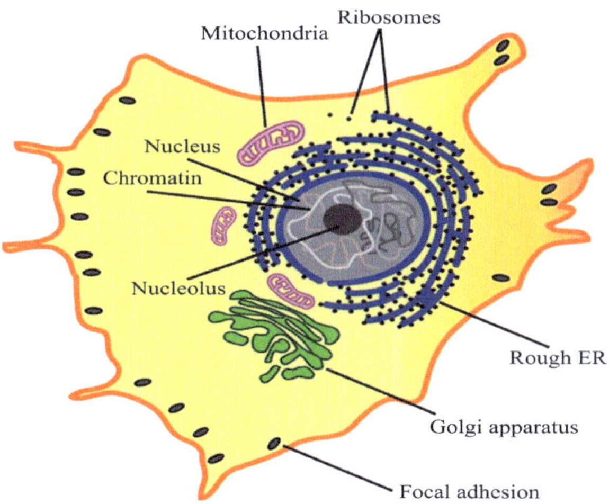


**Figure 2.21.** Within the cell, cytoplasmic organelles.

**Source:** https://www.researchgate.net/figure/Cytoplasmic-organelles-inside-an-eukaryotic-cell_fig1_323043113

Every cell has the same fundamental collection of intracellular organelles that are divided into 2 categories (Knight & Scrutton, 1986):

- The term "membranous organelles" refers to organelles that have membranes that divide their interior environment from the surrounding cytoplasm.
- Organelles that do not have membranes and are non-membranous.

Within the cytoplasm, membranous organelle membranes produce tubular, vesicular, and other structural sequences that can be convoluted (just like in smooth-surfaced endoplasmic reticulum) or plicated (just like in the interior mitochondrial membrane).

Furthermore, every kind of organelle has its collection of proteins (Hartl & Hayer-Hartl, 2002).

Such proteins are either integrated into the membranes of membranous organelles or sequestered inside their lumens. Since its hydrolytic activity will be damaging to the cell, lysosome enzymes have been segregated from the cytoplasmic matrix by a specialized enzyme resistant membrane.

The distinctive proteins in non-membranous organelles normally self-assemble into polymers that compose the cytoskeleton's structural components (Ishihama et al., 2008).

### 2.5.3. Ribosomes

Ribosomes are membrane-bound organelles found in all kinds of cells (eukaryotes and prokaryotes)

Ribosomes are electron-dense particles that are made up of rRNA and proteins.

Because all ribosomes contain 2 subunits of varying sizes that work together to catalyze the procedure of protein translation (protein synthesis), they are more common in protein-secreting cells (Green & Noller, 1997).

Small subunit rRNA molecules are generated in the nucleus of eukaryotic cells. Their many proteins (large subunits) are made in the cytoplasm before entering the nucleus and interacting with rRNAs. The small and big subunits that have been formed exit the nucleus and enter the cytoplasm to engage in the synthesis of proteins.

Ribosomes are frequently connected to the endoplasmic reticulum, but they can also exist freely in the cytoplasm, either individually or as polysomes or polyribosomes (Lafontaine & Tollervey, 2001).

The location of the ribosome in the cell determines the kind of protein that is made by the cell. If the ribosomes are given the freedom to roam about the cell, they will make proteins that will be used by the cell itself. When ribosomes are coupled to the endoplasmic reticulum, proteins synthesized on the rough endoplasmic reticulum are useful both within and outside of the cell (Shine & Dalgarno, 1975).

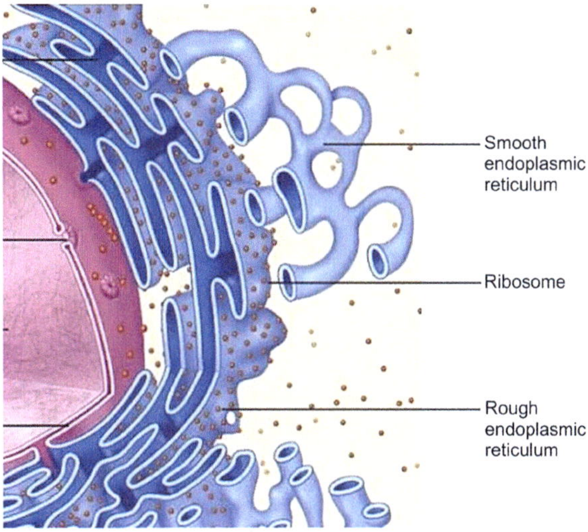


**Figure 2.22.** Ribosome Functions and Structure.

**Source:** https://www.jotscroll.com/forums/3/posts/274/ribosomes-function-what-are-ribosomes-types-ribosome-function-diagram.html

## 2.5.4. Endoplasmic Reticulum (ER)

The ER is a structure of flattened sacs with membrane cisternae spreading across the cytoplasm. It frequently accounts for greater than half of the cell's entire membrane. The existence of a certain form of extensive cytoplasmic structure, then known as the gastroplasty, had been originally discovered in the late nineteenth century while investigations of stained cells revealed the existence of a certain kind of general cytoplasmic structure. In the 1940s, when it was given its current name, the electron microscope made it feasible to investigate the morphology of such organelle (Sundar Rajan et al., 2007).

SER and RER are 2 functionally separate types of the endoplasmic reticulum. The existence of protein-synthesizing particles termed ribosomes connected to the RER's exterior surface distinguishes the 2 morphologically.

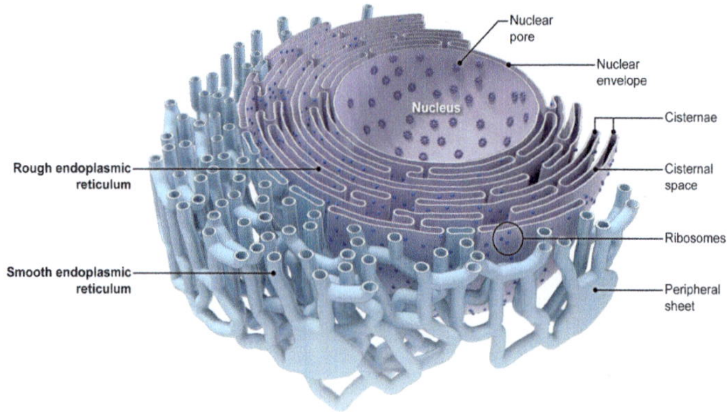


**Figure 2.23.** Endoplasmic reticulum (ER) structure.

**Source:** eLife 2016;5:e20468 *DOI: 10.7554/eLife.20468*

The endoplasmic reticulum is the biggest organelle among most eukaryotic cells (Lynes & Simmen, 2011).

## Membrane Bounded Organelles

This network of intercommunicating passageways and sacs is established through a continuous membrane that encompasses an area known as cisternae. The reticulum is a network that runs from the surface of the nucleus to the membrane of the cell (Bravo et al., 2012).

One of the most important functions of the endoplasmic reticulum is the transfer of materials. This is accomplished by producing transport vesicles that carry big molecules to other sections of the cell. The majority of the time, such vesicles are the route to the plasma membrane or the Golgi apparatus.

RER and SER are the forms of the endoplasmic reticulum that are distinguished by the existence or absence of ribosomes (Zampese et al., 2011).

### 2.5.5. Transmission EM of RER&SER

#### *2.5.5.1. Rough ER (RER)*

A ribosome-bearing membrane is found on the side of the membrane which is exposed to the cytoplasm. Proteins have been generated here and transported into the endoplasmic reticulum, in which they are processed and modified. A few such proteins are integrated into the membrane, while others are exported from the membrane. RER may be present in all cells, except erythrocytes, and is particularly common in pancreatic cells, fibroblasts, and plasma cells, among other tissue (Rost, 2006).

The following are the functions of RER:

- Contributes to the production of a protein that will be exported from the cell.
- Polypeptides that have just been produced are modified.
- Multichain protein assembly
- The glycoprotein's first glycosylation (that means the insertion of glucose into the protein).

The RER is made up of a collection of flattened sacs that are linked. Secretory cells, which specialize in these activities and play a crucial role in glycoprotein and protein creation and export, are the ideal places to look. Secretory cells in the human body include liver cells, which secrete serum proteins such as albumin, endocrine cells, which secrete peptide hormones such as insulin, salivary gland and pancreatic acinar cells, which secrete digestive enzymes, mammary gland cells, which secrete milk proteins, and cartilage cells, which secrete proteoglycans and collagen (Mas et al., 2008).

Proteins are created by ribosomes, which are little particles that need amino acids to do so. They are made up of 4 molecules of RNA and forty to eighty proteins that are organized into small and big subunits. Ribosomes are either unbound (for example, not attached to membranes) or bound to the RER in the cytoplasm of the cell. All of the proteins that are secreted from the cell are produced on membrane-bound ribosomes, which include lysosomal enzymes, Golgi proteins, membrane proteins, and proteins that are destined for the Golgi apparatus (Rost-Roszkowska et al., 2020). Free ribosomes are responsible for the production of proteins in the cytoplasm and those connected to the inner side of the outer membrane, as well as the production of proteins that will be incorporated into the mitochondria, nucleus, peroxisomes, chloroplasts, and other organelles. Proteins have unique characteristics that allow them to be labeled for transport to particular locations both inside and outside the cell (Okouchi et al., 1994). Günter Blobel, a German molecular and cellular biologist, and David Sabatini, an Argentinian cellular biologist, proposed in 1971 that the amino-terminal section of a protein (the very 1st component of the molecule to be synthesized) may operate as a "signal pattern." They hypothesized that this signal pattern will make it easier for the developing protein to connect to the endoplasmic reticulum membrane and guide it into or via membrane into the endoplasmic reticulum lumen (interior) (Horvath et al., 2005).

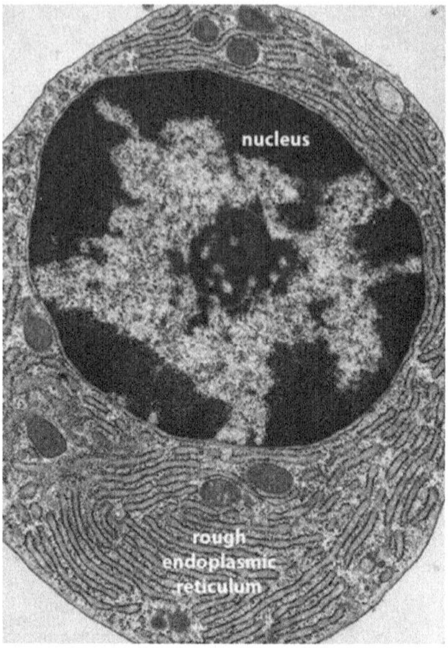


**Figure 2.24.** TEM of RER.

**Source:** Sanvictores T, Davis DD. Histology, Rough Endoplasmic Reticulum. [Updated 2021 Aug 11]. In: StatPearls [Internet]. Treasure Island (FL): StatPearls Publishing; 2022 Jan-. Available from: https://www.ncbi.nlm.nih.gov/books/NBK563126/

A vast body of experimental data is gathered to support the signal idea. On an unbound ribosome, the process of translating the drawing for a particular protein programmed in a messenger RNA molecule gets underway. The signal sequence attaches to a complex of six proteins and one RNA molecule, which is known as the signal recognition particle, in the same way as the developing protein, which contains the signal sequence at its amino-terminal end, develops from the ribosome (Truchan et al., 2016). The signal recognition particle is made up of six proteins and one RNA molecule, which work together to recognize signals. The SRP also attaches to the ribosome, preventing the protein from completing its creation and being released. The endoplasmic reticulum membrane comprises receptor sites that are responsible for binding the SRP-ribosome complex to the RER membrane. As a result of the binding, translation is restarted, with the SRP disentangling from the complicated and the signal sequence and the remaining portion of the nascent protein threading through the membrane and into the endoplasmic reticulum lumen (through a channel known as a translocon (Eng & Youson, 1992). This is the moment at which the protein is irreversibly separated from the cytoplasm. In the majority of situations, the signal pattern is cleaved off the protein through an enzyme known as signal peptidase when the protein emerges from the luminal surface of the endoplasmic reticulum membrane. Additionally, oligosaccharide chains have been frequently attached to proteins via a process known as glycosylation, resulting in the formation of glycoproteins. While in the lumen of the endoplasmic reticulum, the protein folds into the distinctive 3D shape that it is known for (Chajec et al., 2014).

Proteins that would be released by the cell diffuse into the transitional section of the endoplasmic reticulum *which is mainly devoid of ribosomes*, from the lumen. The chemicals are packed into tiny membrane-bound transport vesicles that leave the endoplasmic reticulum *membrane and travel across the cytoplasm to a targeted membrane, generally the complex of Golgi. The membrane of the transport vesicle merges with the membrane of Golgi at this point, and the components of the vesicle are transported inside the Golgi lumen. The sidedness of the membranes is preserved in this procedure, as in all vesicle budding and fusing procedures; such that, the cytoplasmic surface of the membrane constantly faces the outer side, and the luminal composition has always been segregated from the cytoplasm (*Masurovsky et al., 1983).

Some non-secretory proteins produced on the rough endoplasmic reticulum stay in the cell's membrane system. In addition to the signal pattern, such membrane proteins contain one or more anchor regions, which are composed of amino acids that are lipid-soluble. The amino acids tie the protein to the phospholipid bilayer of the Endoplasmic reticulum preventing it from passing entirely into the Endoplasmic reticulum lumen (Zhang & Hua, 2014).

### 2.5.5.2. Smooth Endoplasmic Reticulum (SER)

They have no connected ribosomes and are persistent with RER. SER synthesizes phospholipids found in membranes and performs a variety of different tasks, based on the cell (Callea et al., 1992).

The smooth ER, which is composed of a network of small tubular membrane vesicles, is involved in a diverse spectrum of functions that vary from cell to cell. One of the most important tasks is the production of phospholipids, which are important components of the plasma membrane and the inner membranes of cells. Fatty acids, glycerol phosphate, and other tiny water-soluble molecules are converted into cholesterol and phospholipids by enzymes linked to the endoplasmic reticulum membrane (ER membrane) with their active sites facing the cytoplasm (Tsukita & Ishikawa, 1976). Several phospholipids remain in the endoplasmic reticulum, where they can "flip" from the cytoplasmic side of the bilayer, where they were formed, to the axoplasmic side of the bilayer, where they were formed, thanks to the action of enzymes found inside the membranes. The endoplasmic reticulum membrane grows symmetrically as a result of this mechanism. Special phospholipid transfer proteins transmit other phospholipids by the cytoplasm to certain other membranous structures like the mitochondrion and the cell membrane (Ebner et al., 2008).

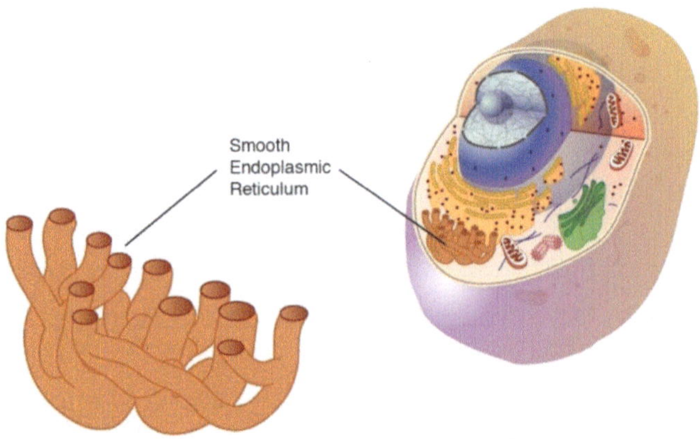


**Figure 2.25.** Graphical depiction of SER.

**Source:** https://www.genome.gov/genetics-glossary/Endoplasmic-Reticulum-Smooth

The smooth ER in liver cells is responsible for detoxifying a broad range of chemicals generated via metabolic activities. The cytochrome P450 enzymes found in the liver smooth ER catalyze the degradation of carcinogens and other chemical compounds. Cholesterol is altered in the SER in cells of the gonads and adrenal glands during one stage of its conversion to steroid hormones. Ultimately, the sarcoplasmic reticulum, or smooth ER, in muscle cells sequesters calcium ions from the cytoplasm. Calcium ions are produced whenever the muscle is stimulated via nerve impulses, triggering muscular contraction (Walz, 1982).

## 2.5.6. Golgi Apparatus

Secretory glycoproteins and proteins are changed, finished, and exported at the Golgi complex. The organelle was first found in 1898 by Italian cytologist Camillo Golgi, and it has a characteristic morphology consisting of 5 to 8 flattened, disk-shaped, membrane-defined cisternae piled in a stack (Grond et al., 2020). Secretory glycoproteins and proteins, cell membrane glycoproteins and proteins, lysosomal proteins, and certain glycolipids all pass through the Golgi complex at some point during their development. The Golgi transports the majority of the cell wall material in plant cells (Farquhar & Palade, 1998).

With a "trans" face at the cell membrane, a "cis" face in the RER transitional area, and a middle part, the Golgi apparatus is structurally polarised. Each portion of such a face is biochemically distinct, with a distinct enzyme makeup. In comparison to the others, the membranes on the cis face are generally thinner (Glick & Malhotra, 1998).

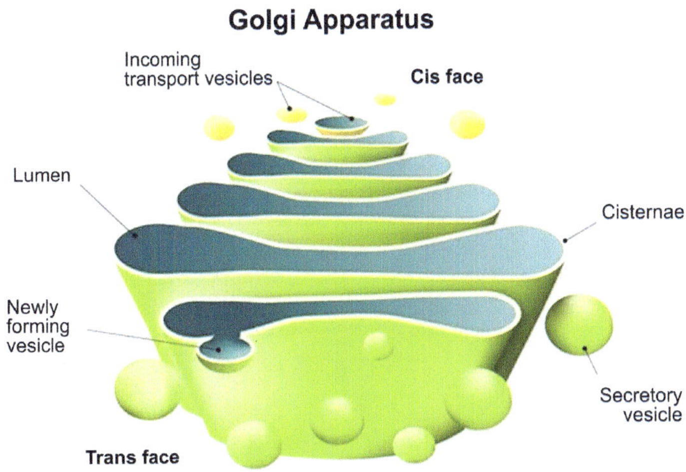


**Figure 2.26.** The Golgi apparatus. The Golgi Complex is vital for protein transport and modification inside the cell.

**Source:** https://www.news-medical.net/life-sciences/The-Endoplasmic-Reticulum-and-Golgi-Body-Whats-the-Difference.aspx

Certain chemical changes can occur as secretory proteins pass through the Golgi. Carbohydrate group modification is one of the most essential. Several secretory proteins have been glycosylated in the endoplasmic reticulum, as previously mentioned. Particular enzymes in the Golgi remove particular mannose residues from glycoprotein oligosaccharide chains and replace them with other sugars like sialic acid and galactose (Lipsky & Pagano, 1985). Glycosidases and glycosyltransferases are the collective names for such enzymes. Certain secretory proteins might stop being transported if their carbohydrate groups were incorrectly altered or not permitted to mature. Carbohydrate groups are essential for protein stability or function in some cases, as well as for the molecule to be targeted to a specific place (Munro, 1998).

Within the Golgi or secretory vesicles are proteases that break certain secretory proteins at specific amino acid positions. When inactive proinsulin is transformed into active insulin by removing a sequence of amino acids, this typically results in secretory protein activation (Mollenhauer & Whaley, 1963).

## REFERENCES


1. Allfrey, V. G., Faulkner, R., & Mirsky, A. (1964). Acetylation and methylation of histones and their possible role in the regulation of RNA synthesis. Proceedings of the National Academy of Sciences of the United States of America, 51(5), pp. 786.
2. Aronson, P. S. (1985). Kinetic properties of the plasma membrane Na+-H+ exchanger. Annual review of physiology, 47(1), 545-560.
3. Balhorn, R., Corzett, M., Allen, M. J., Lee, C. S., Barbee Jr, T. W., Koch, J. A., ... & Attwood Jr, D. T. (1993, January). Application of x rays to the analysis of DNA packaging in mammalian sperm. In Soft X-Ray Microscopy (Vol. 1741, pp. 374-385). SPIE.
4. Basheer, L., Kropp, J., & Cleland, D. J. (2001). Assessment of the durability of concrete from its permeation properties: a review. Construction and building materials, 15(2-3), 93-103.
5. Basheer, P. M. (2001). Permeation analysis. Handbook of Analytical Techniques in Concrete Science and Technology: Principles, Techniques and Applications, Editors VS Ramachandran and JJ Beaudoin, Noyes Publications, (Vol. 1, pp. 658-727).
6. Battle, A., Mostafavi, S., Zhu, X., Potash, J. B., Weissman, M. M., McCormick, C., ... & Koller, D. (2014). Characterizing the genetic basis of transcriptome diversity through RNA-sequencing of 922 individuals. Genome research, 24(1), 14-24.
7. Beney, L., & Gervais, P. (2001). Influence of the fluidity of the membrane on the response of microorganisms to environmental stresses. Applied microbiology and biotechnology, 57(1), 34-42.
8. Bentley, D. (1999). Coupling RNA polymerase II transcription with pre-mRNA processing. Current opinion in cell biology, 11(3), 347-351.
9. Bernt, K. M., & Armstrong, S. A. (2011). Targeting epigenetic programs in MLL-rearranged leukemias. Hematology 2010, the American Society of Hematology Education Program Book, 2011(1), 354-360.
10. Bétermier, M., & Duharcourt, S. (2014). Programmed rearrangement in ciliates: Paramecium. Microbiology spectrum, 2(6), 2-6.
11. Bock, C. T., Schranz, P., Schröder, C. H., & Zentgraf, H. (1994). Hepatitis B virus genome is organized into nucleosomes in the nucleus of the infected cell. Virus genes, 8(2), 215-229.
12. Bowie, J. U. (2001). Stabilizing membrane proteins. Current opinion in structural biology, 11(4), 397-402.

13. Branton, D., & Deamer, D. W. (1972). Membrane structure. In Membrane Structure (Vol. 1, pp. 1-70). Springer, Vienna.
14. Bravo, R., Gutierrez, T., Paredes, F., Gatica, D., Rodriguez, A. E., Pedrozo, Z., ... & Lavandero, S. (2012). Endoplasmic reticulum: ER stress regulates mitochondrial bioenergetics. The international journal of biochemistry & cell biology, 44(1), 16-20.
15. Buckhout-White, S., Person, C., Medintz, I. L., & Goldman, E. R. (2018). Restriction enzymes as a target for DNA-based sensing and structural rearrangement. ACS omega, 3(1), 495-502.
16. Burke, B., & Stewart, C. L. (2002). Life at the edge: the nuclear envelope and human disease. Nature reviews Molecular cell biology, 3(8), 575-585.
17. Callea, F., Brisiptti, M., Fabbretti, G., Bonino, F., & Desmet, V. J. (1992). Hepatic endoplasmic reticulum storage diseases. Liver, 12(6), 357-362.
18. Chajec, Ł., Rost-Roszkowska, M. M., Vilimova, J., & Sosinka, A. (2012). Ultrastructure and regeneration of midgut epithelial cells in *Lithobius forficatus* (Chilopoda, Lithobiidae). Invertebrate Biology, 131(2), 119-132.
19. Chajec, Ł., Sonakowska, L., & Rost-Roszkowska, M. M. (2014). The fine structure of the midgut epithelium in a centipede, *Scolopendra cingulata* (Chilopoda, Scolopendridae), with the special emphasis on epithelial regeneration. Arthropod Structure & Development, 43(1), 27-42.
20. Chen, Y. K., & Li, K. B. (2013). Predicting membrane protein types by incorporating protein topology, domains, signal peptides, and physicochemical properties into the general form of Chou's pseudo amino acid composition. Journal of Theoretical Biology, 318(1), 1-12.
21. Chothia, C., & Jones, E. Y. (1997). The molecular structure of cell adhesion molecules. Annual review of biochemistry, 66(1), 823-862.
22. Cirillo, V. P. (1962). Mechanism of glucose transport across the yeast cell membrane. Journal of bacteriology, 84(3), 485-491.
23. Cooper, K., Jakobsson, E., & Wolynes, P. (1985). The theory of ion transport through membrane channels. Progress in biophysics and molecular biology, 46(1), 51-96.
24. Creutz, C. E. (1992). The annexins and exocytosis. Science, 258(5084), 924-931.

25. Crick, F. H. C., & Hughes, A. F. W. (1950). The physical properties of cytoplasm. Experimental cell research, 1(1), 37-80.
26. Cronan Jr, J. E., & Gelmann, E. P. (1975). Physical properties of membrane lipids: biological relevance and regulation. Bacteriological reviews, 39(3), 232-256.
27. Cronan, J. E. (2003). Bacterial membrane lipids: where do we stand?. Annual reviews in microbiology, 57(1), 203-224.
28. Cross, B., Sinning, I., Luirink, J., & High, S. (2009). Delivering proteins for export from the cytosol. Nature reviews Molecular cell biology, 10(4), 255-264.
29. Culbertson, M. R. (1999). RNA surveillance: unforeseen consequences for gene expression, inherited genetic disorders and cancer. Trends in Genetics, 15(2), 74-80.
30. Dellago, C., & Hummer, G. (2006). Kinetics and mechanism of proton transport across membrane nanopores. Physical review letters, 97(24), 245901.
31. Dell'Angelica, E. C., Mullins, C., Caplan, S., & Bonifacino, J. S. (2000). Lysosome-related organelles. The FASEB Journal, 14(10), 1265-1278.
32. Devuyst, O., & Rippe, B. (2014). Water transport across the peritoneal membrane. Kidney International, 85(4), 750-758.
33. Dillon, N. (2006). Gene regulation and large-scale chromatin organization in the nucleus. Chromosome Research, 14(1), 117-126.
34. Ebner, T., Moser, M., Shebl, O., Sommerguber, M., & Tews, G. (2008). Prognosis of oocytes showing aggregation of smooth endoplasmic reticulum. Reproductive biomedicine online, 16(1), 113-118.
35. Eng, F., & Youson, J. H. (1992). Morphology of the bile ducts of the brook lamprey, *Lampetra lamottenii* (Le Sueur) before and during infection with the nematode, *Truttaedacnitis stelmioides* (Vessichelli, 1910)(Nematoda: Cucullanidae). The Anatomical Record, 234(2), 201-214.
36. Ernst, R., Ejsing, C. S., & Antonny, B. (2016). Homeoviscous adaptation and the regulation of membrane lipids. Journal of molecular biology, 428(24), 4776-4791.
37. Fang, W., Wang, X., Bracht, J. R., Nowacki, M., & Landweber, L. F. (2012). Piwi-interacting RNAs protect DNA against loss during Oxytricha genome rearrangement. Cell, 151(6), 1243-1255.

38. Farquhar, M. G., & Palade, G. E. (1998). The Golgi apparatus: 100 years of progress and controversy. Trends in cell biology, 8(1), 2-10.
39. Field, L. D., Delehanty, J. B., Chen, Y., & Medintz, I. L. (2015). Peptides for specifically targeting nanoparticles to cellular organelles: quo vadis?. Accounts of chemical research, 48(5), 1380-1390.
40. Fischer, D., Li, Y., Ahlemeyer, B., Krieglstein, J., & Kissel, T. (2003). In vitro cytotoxicity testing of polycations: influence of polymer structure on cell viability and hemolysis. Biomaterials, 24(7), 1121-1131.
41. Frey, T. G., Perkins, G. A., & Ellisman, M. H. (2006). Electron tomography of membrane-bound cellular organelles. Annu. Rev. Biophys. Biomol. Struct., 35(1), 199-224.
42. Friend, D. R. (1992). In vitro skin permeation techniques. Journal of Controlled Release, 18(3), 235-248.
43. Gerace, L., & Blobel, G. (1982, January). Nuclear lamina and the structural organization of the nuclear envelope. In Cold Spring Harbor Symposia on Quantitative Biology (Vol. 46, pp. 967-978). Cold Spring Harbor Laboratory Press.
44. Gietl, C. (1992). Malate dehydrogenase isoenzymes: cellular locations and role in the flow of metabolites between the cytoplasm and cell organelles. Biochimica Et Biophysica Acta (BBA)-Bioenergetics, 1100(3), 217-234.
45. Glaumann, H., Ericsson, J. L., & Marzella, L. (1981). Mechanisms of intralysosomal degradation with special reference to autophagocytosis and heterophagocytosis of cell organelles. International review of cytology, 73(1), 149-182.
46. Glick, B. S., & Malhotra, V. (1998). The curious status of the Golgi apparatus. Cell, 95(7), 883-889.
47. Görlich, D., & Kutay, U. (1999). Transport between the cell nucleus and the cytoplasm. Annual review of cell and developmental biology, 15(1), 607-660.
48. Green, R., & Noller, H. F. (1997). Ribosomes and translation. Annual review of biochemistry, 66(1), 679-716.
49. Grond, R., Veenendaal, T., Duran, J. M., Raote, I., Van Es, J. H., Corstjens, S., ... & Rabouille, C. (2020). The function of GORASPs in Golgi apparatus organization in vivo. Journal of Cell Biology, 219(9).
50. Guidotti, G. (1972). Membrane proteins. Annual review of biochemistry, 41(1), 731-752.

51. Håkansson, S., Charron, A. J., & Sibley, L. D. (2001). Toxoplasma evacuoles: a two-step process of secretion and fusion forms the parasitophorous vacuole. The EMBO journal, 20(12), 3132-3144.
52. Hanany, A., Mekareeya, N., & Zaffaroni, A. (2008). Partition functions for membrane theories. Journal of High Energy Physics, 2008(09), (Vol. 1, pp. 90).
53. Harada, S. (2005). The broad anti-viral agent glycyrrhizin directly modulates the fluidity of plasma membrane and HIV-1 envelope. Biochemical Journal, 392(1), 191-199.
54. Hartl, F. U., & Hayer-Hartl, M. (2002). Molecular chaperones in the cytosol: from nascent chain to folded protein. Science, 295(5561), 1852-1858.
55. Horvath, E., Kovacs, K., Smyth, H. S., Cusimano, M., & Singer, W. (2005). Silent adenoma subtype 3 of the pituitary—immunohistochemical and ultrastructural classification: a review of 29 cases. Ultrastructural pathology, 29(6), 511-524.
56. Ishihama, Y., Schmidt, T., Rappsilber, J., Mann, M., Hartl, F. U., Kerner, M. J., & Frishman, D. (2008). Protein abundance profiling of the *Escherichia coli* cytosol. BMC genomics, 9(1), 1-17.
57. Iwamoto, K., Bundo, M., & Kato, T. (2009). Serotonin receptor 2C and mental disorders: genetic, expression, and RNA editing studies. RNA biology, 6(3), 248-253.
58. Kao, C. C., Singh, P., & Ecker, D. J. (2001). De novo initiation of viral RNA-dependent RNA synthesis. Virology, 287(2), 251-260.
59. Khalili-Araghi, F., Gumbart, J., Wen, P. C., Sotomayor, M., Tajkhorshid, E., & Schulten, K. (2009). Molecular dynamics simulations of membrane channels and transporters. Current opinion in structural biology, 19(2), 128-137.
60. Knight, D. E., & Scrutton, M. C. (1986). Gaining access to the cytosol: the technique and some applications of electropermeabilization. Biochemical Journal, 234(3), 497.
61. Kotnik, T., & Miklavčič, D. (2006). Theoretical evaluation of voltage inducement on internal membranes of biological cells exposed to electric fields. Biophysical Journal, 90(2), 480-491.
62. Krieg, P. A., & Melton, D. A. (1987). [25] In vitro RNA synthesis with SP6 RNA polymerase. In Methods in enzymology (Vol. 155, pp. 397-415). Academic Press.

63. Kumar, V., Westra, H. J., Karjalainen, J., Zhernakova, D. V., Esko, T., Hrdlickova, B., ... & Wijmenga, C. (2013). Human disease-associated genetic variation impacts large intergenic non-coding RNA expression. PLoS genetics, 9(1), pp. 1-6.

64. Lafontaine, D. L., & Tollervey, D. (2001). The function and synthesis of ribosomes. Nature Reviews Molecular Cell Biology, 2(7), 514-520.

65. Levin-Klein, R., & Bergman, Y. (2014). Epigenetic regulation of monoallelic rearrangement (allelic exclusion) of antigen receptor genes. Frontiers in Immunology, 5(1), pp. 625.

66. Li, G., & Zhu, P. (2015). Structure and organization of chromatin fiber in the nucleus. FEBS Letters, 589(20), 2893-2904.

67. Lindahl, T. (1993). Instability and decay of the primary structure of DNA. nature, 362(6422), 709-715.

68. Lipsky, N. G., & Pagano, R. E. (1985). A vital stain for the Golgi apparatus. Science, 228(4700), 745-747.

69. Lynes, E. M., & Simmen, T. (2011). Urban planning of the endoplasmic reticulum (ER): how diverse mechanisms segregate the many functions of the ER. Biochimica Et Biophysica Acta (BBA)-Molecular Cell Research, 1813(10), 1893-1905.

70. Martin, G. G., Hose, J. E., & Kim, J. J. (1987). Structure of hematopoietic nodules in the ridgeback prawn, *Sicyonia ingentis*: light and electron microscopic observations. Journal of morphology, 192(3), 193-204.

71. Mas, N., Tasci, I., Comert, B., Ocal, R., & Mas, M. R. (2008). Ursodeoxycholic acid treatment improves hepatocyte ultrastructure in rat liver fibrosis. World Journal of Gastroenterology: WJG, 14(7), 1108.

72. Masurovsky, E. B., Peterson, E. R., Crain, S. M., & Horwitz, S. B. (1983). Morphological alterations in dorsal root ganglion neurons and supporting cells of organotypic mouse spinal cord-ganglion cultures exposed to taxol. Neuroscience, 10(2), 491-509.

73. Maurel, C., Verdoucq, L., Luu, D. T., & Santoni, V. (2008). Plant aquaporins: membrane channels with multiple integrated functions. Annu. Rev. Plant Biol., 59(1), 595-624.

74. Mohandas, N., & Gallagher, P. G. (2008). Red cell membrane: past, present, and future. Blood, The Journal of the American Society of Hematology, 112(10), 3939-3948.

75. Molino, D., Zemirli, N., Codogno, P., & Morel, E. (2017). The

journey of the autophagosome through mammalian cell organelles and membranes. Journal of molecular biology, 429(4), 497-514.

76. Mollenhauer, H. H., & Whaley, W. G. (1963). An observation on the functioning of the Golgi apparatus. The Journal of cell biology, 17(1), 222.

77. Monneron, A., & Bernhard, W. (1969). Fine structural organization of the interphase nucleus in some mammalian cells. Journal of ultrastructure research, 27(3-4), 266-288.

78. Montanaro, L., Treré, D., & Derenzini, M. (2008). Nucleolus, ribosomes, and cancer. The American journal of pathology, 173(2), 301-310.

79. Müller-McNicoll, M., Botti, V., de Jesus Domingues, A. M., Brandl, H., Schwich, O. D., Steiner, M. C., ... & Neugebauer, K. M. (2016). SR proteins are NXF1 adaptors that link alternative RNA processing to mRNA export. Genes & development, 30(5), 553-566.

80. Müller-Plathe, F. (1994). Permeation of polymers—a computational approach. Acta Polymerica, 45(4), 259-293.

81. Munro, S. (1998). Localization of proteins to the Golgi apparatus. Trends in cell biology, 8(1), 11-15.

82. Neidhardt, F. C. (1964). The regulation of RNA synthesis in bacteria. Progress in nucleic acid research and molecular biology, Vol. 3, 145-181.

83. Novak, I. L., Kraikivski, P., & Slepchenko, B. M. (2009). Diffusion in cytoplasm: effects of excluded volume due to internal membranes and cytoskeletal structures. Biophysical Journal, 97(3), 758-767.

84. O'Brien, R. M., & Granner, D. K. (1996). Regulation of gene expression by insulin. Physiological reviews, 76(4), 1109-1161.

85. Okouchi, Y., Sasaki, K., & Tamaki, T. (1994). Ultrastructural changes in hepatocytes, sinusoidal endothelial cells and macrophages in hypothermic preservation of the rat liver with University of Wisconsin solution. Virchows Archiv, 424(5), 477-484.

86. Parpart, A. K., & Dziemian, A. J. (1940, January). The chemical composition of the red cell membrane. In Cold Spring Harbor Symposia on Quantitative Biology (Vol. 8, pp. 17-24). Cold Spring Harbor Laboratory Press.

87. Pérez, C., Leng, M., & Malinge, J. M. (1997). Rearrangement of interstrand cross-links into intrastrand cross-links in cis-

diamminedichloroplatinum (II)-modified DNA. Nucleic acids research, 25(4), 896-903.

88. Pohorille, A., Schweighofer, K., & Wilson, M. A. (2005). The origin and early evolution of membrane channels. Astrobiology, 5(1), 1-17.
89. Postnov, Y. V., & Orlov, S. N. (1985). Ion transport across plasma membrane in primary hypertension. Physiological reviews, 65(4), 904-945.
90. Pucadyil, T. J., & Schmid, S. L. (2009). Conserved functions of membrane active GTPases in coated vesicle formation. Science, 325(5945), 1217-1220.
91. Radons, J., & Multhoff, G. (2005). Immunostimulatory functions of membrane-bound and exported heat shock protein 70. Exerc Immunol Rev, 11(1), 17-33.
92. Rapoport, T. A., Rolls, M. M., & Jungnickel, B. (1996). Approaching the mechanism of protein transport across the ER membrane. Current opinion in cell biology, 8(4), 499-504.
93. Ren, Q., Chen, K., & Paulsen, I. T. (2007). TransportDB: a comprehensive database resource for cytoplasmic membrane transport systems and outer membrane channels. Nucleic acids research, 35(suppl_1), D274-D279, (Vol. 2, pp. 1-3).
94. Rogers, C. E. (1985). Permeation of gases and vapours in polymers. In Polymer permeability (Vol.1, pp. 11-73). Springer, Dordrecht.
95. Romano, M., Ruggiero, A., Squeglia, F., Maga, G., & Berisio, R. (2020). A structural view of SARS-CoV-2 RNA replication machinery: RNA synthesis, proofreading and final capping. Cells, 9(5), 1267.
96. Rost, M. M. (2006). Comparative Studies on Regeneration of the Midgut Epithelium in Lepisma saccharina and Thermobia domestica. Annals of the Entomological Society of America, 99(5), 910-916.
97. Rost-Roszkowska, M. M., Vilimova, J., & Chajec, Ł. (2010). Fine structure of the midgut epithelium in *Atelura formicaria* (Hexapoda, Zygentoma, Ateluridae), with special reference to its regeneration and degeneration. Zool Stud, 49(1), 10-18.
98. Rost-Roszkowska, M., Poprawa, I., Chajec, Ł., Chachulska-Żymełka, A., Leśniewska, M., & Student, S. (2020). Effects of short-and long-term exposure to cadmium on salivary glands and fat body of soil centipede *Lithobius forficatus* (Myriapoda, Chilopoda): Histology and ultrastructure. Micron, 137(1), 102915.

99. Roux, B., & Schulten, K. (2004). Computational studies of membrane channels. Structure, 12(8), 1343-1351.

100. Sahay, G., Alakhova, D. Y., & Kabanov, A. V. (2010). Endocytosis of nanomedicines. Journal of controlled release, 145(3), 182-195.

101. Sanders, M. M. (1978). Fractionation of nucleosomes by salt elution from micrococcal nuclease-digested nuclei. The Journal of cell biology, 79(1), 97-109.

102. Shibata, Y., Hu, J., Kozlov, M. M., & Rapoport, T. A. (2009). Mechanisms shaping the membranes of cellular organelles. Annual Review of Cell and Developmental, 25(1), 329-354.

103. Shine, J., & Dalgarno, L. (1975). Determinant of cistron specificity in bacterial ribosomes. Nature, 254(5495), 34-38.

104. Silverstein, S. C., Steinman, R. M., & Cohn, Z. A. (1977). Endocytosis. Annual review of biochemistry, 46(1), 669-722.

105. Singh, P., Alley, T. L., Wright, S. M., Kamdar, S., Schott, W., Wilpan, R. Y., ... & Graber, J. H. (2009). Global changes in processing of mRNA 3′ untranslated regions characterize clinically distinct cancer subtypes. Cancer Research, 69(24), 9422-9430.

106. Sola, I., Almazan, F., Zúñiga, S., & Enjuanes, L. (2015). Continuous and discontinuous RNA synthesis in coronaviruses. Annual review of virology, 2(1), 265-288.

107. Somosy, Z. (2000). Radiation response of cell organelles. Micron, 31(2), 165-181.

108. Sosunov, V., Sosunova, E., Mustaev, A., Bass, I., Nikiforov, V., & Goldfarb, A. (2003). Unified two-metal mechanism of RNA synthesis and degradation by RNA polymerase. The EMBO journal, 22(9), 2234-2244.

109. Sprott, G. D. (1992). Structures of archaebacterial membrane lipids. Journal of bioenergetics and biomembranes, 24(6), 555-566.

110. Steven Ward, W., & Coffey, D. S. (1991). DNA packaging and organization in mammalian spermatozoa: comparison with somatic cell. Biology of reproduction, 44(4), 569-574.

111. Sun, G. Y., & Sun, A. Y. (1985). Ethanol and membrane lipids. Alcoholism, clinical and experimental research, 9(2), 164-180.

112. Sundar Rajan, S., Srinivasan, V., Balasubramanyam, M., & Tatu, U. (2007). Endoplasmic reticulum (ER) stress & diabetes. The Indian journal of medical research, 125(3), 411-24.

113. Truchan, H. K., Cockburn, C. L., Hebert, K. S., Magunda, F., Noh, S. M., & Carlyon, J. A. (2016). The pathogen-occupied vacuoles of *Anaplasma phagocytophilum* and *Anaplasma marginale* interact with the endoplasmic reticulum. Frontiers in cellular and infection microbiology, (Vol. 6, pp. 22).
114. TSUKITA, S., & Ishikawa, H. (1976). Three-dimensional distributionof smooth endoplasmic reticulum in myelinated axons. Microscopy, 25(3), 141-149.
115. Uittenbogaard, A., Everson, W. V., Matveev, S. V., & Smart, E. J. (2002). Cholesteryl ester is transported from caveolae to internal membranes as part of a caveolin-annexin II lipid-protein complex. Journal of Biological Chemistry, 277(7), 4925-4931.
116. Van Meer, G., Voelker, D. R., & Feigenson, G. W. (2008). Membrane lipids: where they are and how they behave. Nature reviews Molecular cell biology, 9(2), 112-124.
117. Vigil, E. L. (1970). Cytochemical and developmental changes in microbodies (glyoxysomes) and related organelles of castor bean endosperm. The Journal of cell biology, 46(3), 435-454.
118. Visser, W. F., Van Roermund, C. W., Ijlst, L., Waterham, H. R., & Wanders, R. J. (2007). Metabolite transport across the peroxisomal membrane. Biochemical Journal, 401(2), 365-375.
119. Walz, B. (1982). Calcium-sequestering smooth endoplasmic reticulum in retinula cells of the blowfly. Journal of ultrastructure research, 81(2), 240-248.
120. Ward, W. S., & Coffey, D. S. (1991). DNA packaging and organization in mammalian spermatozoa: comparison with somatic cells. Biol Reprod, 44(4), 569-74.
121. Watson, J. D., & Crick, F. H. (1953, January). The structure of DNA. In Cold Spring Harbor symposia on quantitative biology (Vol. 18, pp. 123-131). Cold Spring Harbor Laboratory Press.
122. Webster, M., Witkin, K. L., & Cohen-Fix, O. (2009). Sizing up the nucleus: nuclear shape, size and nuclear-envelope assembly. Journal of cell science, 122(10), 1477-1486.
123. Wilschut, A., Wil, F., Robinson, P. J., & McKone, T. E. (1995). Estimating skin permeation. The validation of five mathematical skin permeation models. Chemosphere, 30(7), 1275-1296.
124. Wilson, M. A., & Pohorille, A. (1996). Mechanism of unassisted ion

transport across membrane bilayers. Journal of the American Chemical Society, 118(28), 6580-6587.

125. Wolter, K. G., Hsu, Y. T., Smith, C. L., Nechushtan, A., Xi, X. G., & Youle, R. J. (1997). Movement of Bax from the cytosol to mitochondria during apoptosis. The Journal of cell biology, 139(5), 1281-1292.
126. Xiong, T. C., Bourque, S., Lecourieux, D., Amelot, N., Grat, S., Brière, C., ... & Ranjeva, R. (2006). Calcium signaling in plant cell organelles delimited by a double membrane. Biochimica et Biophysica Acta (BBA)-Molecular Cell Research, 1763(11), 1209-1215.
127. Yeagle, P. L. (1985). Cholesterol and the cell membrane. Biochimica et Biophysica Acta (BBA)-Reviews on Biomembranes, 822(3-4), 267-287.
128. Zampese, E., Fasolato, C., Kipanyula, M. J., Bortolozzi, M., Pozzan, T., & Pizzo, P. (2011). Presenilin 2 modulates endoplasmic reticulum (ER)–mitochondria interactions and Ca2+ cross-talk. Proceedings of the National Academy of Sciences, 108(7), 2777-2782.
129. Zhang, G., & Hua, B. (2014). Fine structure of the midgut of Sinopanorpa tincta (Navás)(Mecoptera: Panorpidae). Tissue and Cell, 46(5), 388-396.

CHAPTER 3

# PROPERTIES OF PROKARYOTIC AND EUKARYOTIC CELLS

CONTENTS

## 3.1. INTRODUCTION

We'll discover how a cell is in this section, as well as what a prokaryotic or eukaryotic cell is. Robert Hooke utilizes a rudimentary composite microscope to preview the cell for the very first time in 1665. He noticed extremely thin cork slices and a plethora of small structures which he compared to monk-like walled chambers. As a result, they were given the term cells. Micrographia published Hooke's explanation of these cells (Cavalier-Smith, 1975). The cells are the building blocks of a biological system, with dimensions ranging from 1 to 100 micrometers. They come in a range of forms and sizes to serve some purposes. Comprehending the form and composition of a cell is essential for knowing the form and composition of a live creature. The capacity to split is one of the most significant qualities of a cell. A cell's presence shows that it would have developed from an earlier cell, and it may also give birth to a new cell. Theodor Schwann is the one to say this. The cell hypothesis arose from pioneering studies on cells by Theodor Schwann and Matthias Jakob Schleiden (Boiteux et al., 2017).

According to their idea,

All organisms are composed of cells.

Life's fundamental building blocks are cells.

Old cells divide into two to form new cells (Nayak et al., 1985).

Rudolf Virchow completed the basic cell theory in 1855 by adding a point to the concept as well as concluding that almost all cells arise from already known cells. The cell idea applies to all organisms, no matter how large or little, simple or complicated they are. The cell hypothesis does not apply to viruses. All organisms have cells, which contain data about all kinds of life. Scientists may examine cells to understand how organisms function since all cells arise from pre-existing cells. We can understand all sorts of living organisms by studying cells and also how they work (Sacchetti et al., 2001).

Blue-green algae, bacteria, plants, and mammals all contain cell structures and might include one or more cells. Unicellular creatures are creatures that have just one cell in their structure (bacteria, blue-green algae, Protozoa, few algae, etc.). Complex creatures are creatures with a large number of cells in their body (fungi, most plants, and animals). Only one kind of cell may exist in a live organism: A. prokaryotic cells; B. eukaryotic cells. Hans Ris suggested the names prokaryotic and eukaryotic in the 1960s. This grouping depends on what way complex they are. Plant and animal cells vary in various ways depending on which kingdom they belong to, such as the animals and plant kingdom. In the next parts, they will be examined in further depth (Krämer, 1996).

## 3.2. PROKARYOTIC CELLS

In Greek, prokaryote signifies "before the nucleus." They encompass all cells that do not have a nucleus and other organelles that are membrane-bound. Prokaryotes include viruses, mycoplasma, bacteria, blue-green algae, or cyanobacteria (Fröhlich & König, 2000),

The majority of prokaryotes are from 1 to 10 meters in length, however, they may be varied in diameter from 0.2 to 750 meters (*Thiomargarita namibiensis*). They are classified into two taxonomic disciplines: bacteria and archaea. The majority of prokaryotes are unicellular, apart from myxobacteria, that undergo multicellular phases during their entire lifespans. They are primarily unicellular organisms with membrane-bound organelles that do not have interior membrane-bound organelles. A normal prokaryotic cell is shown graphically in fig1 (Oikonomou et al., 2016). While prokaryotes have no cell organelles, they do include several inner assemblies like cytoskeletons and ribosomes, that are responsible for the translation of mRNA to proteins.

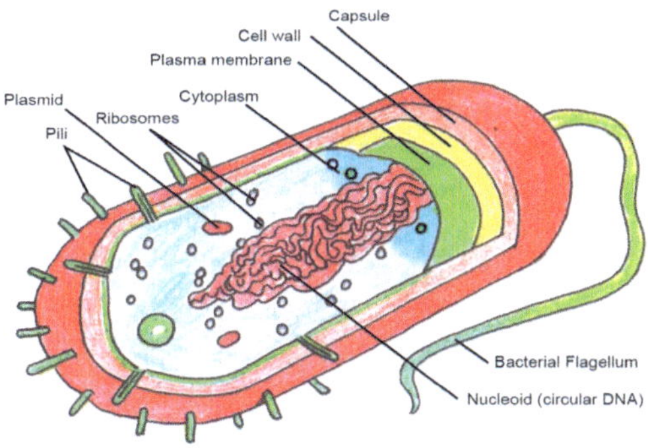


**Figure 3.1** Diagrammatic representation of a prokaryotic cell.

**Source:** https://www.jotscroll.com/forums/3/posts/220/prokaryotic-cells-parts-prokaryotes-definition-examples-diagram.html

Membranous organelles, like vacuoles or membrane systems specialized for certain metabolic functions, like photosynthesis or chemolithotrophy, are recognized in several classes of prokaryotes. Additionally, several species include protein-enclosed microcompartments that serve specific physiological functions (carboxysomes or gas vacuoles) (Schippers et al., 2005).

Individual structures represented in the Figure above are as such, with specifics detailed in subsequent chapters (Apirion & Miczak, 1993):

### 3.2.1. Flagella

It is a lengthy, whip-like protrusion present on the surface of the majority of prokaryotes that assists in cell movement. Apart from its primary role of movement, it often serves as a sensory organelle, becoming susceptible to elements and external temperatures. Numerous Gram-positive and Gram-negative bacteria, Archaea, and certain eukaryotic cells all possess a flagellum ('whip' in Latin). The flagellum is essentially an organelle of mobility that permits motion as well as chemotaxis (Moens & Vanderleyden, 1996). Bacteria may have a single flagellum or multiple, and they could be polarized (having one or more flagella at a single location) or peritrichous (having one or more flagella at a single location) (some of the flagella all over the bacterium). Apart from mobility, flagella perform a variety of different tasks that vary according to bacteria and stage of the bacterial life cycle: for instance, a flagellum may assist in biofilm formation, protein export, and adhesion (Machin, 1958). This section will primarily concentrate on the adhesive capabilities of flagella, with an emphasis on research examining the direct contact of flagella with host targets, but flagellar mobility is frequently a need for adherence and/or invasion. Duan et al. have described the diverse functions of bacterial flagella in pathogenicity in more depth. Prüss et al. and Conrad provide overviews of the function of flagella in biofilms (Macnab, 2003).

The normal flagellum of Escherichia coli is around 10 m in length and 20 nm in diameter. The flagellum requires around 60 structurally and regulating proteins for formation and functioning. The flagellum is composed of cytoplasmic spread machinery, a basal body buried in the cell film (CM), a hook connecting the flat body to the filament, and a propeller filament (Berg, 2003). Flagellar assembly begins along with CM-associated elements of the basal body and the secretory equipment, wherein the rest of the basal body elements, followed by the hook and hook-filament junction proteins, are released. Just after the hook (120 FlgE molecules) has been constructed and the filament capping proteins have indeed been placed, filament formation begins (Silverman & Simon, 1977). The filament is made of around

20,000 flagellins (FliC) proteins which are integrated underneath the distal pentameric FliD cap, which acts as a plug as well as being essential for nascent monomeric flagellin formation. Flagellar rotation is mediated by 2 different parts signaling cascade including chemotaxis-related proteins and the basal body, which works as a rotor (Gibbons, 1981).

### 3.2.2. Capsule

The capsule is present in certain bacterial cells; this additional external layer shields the cell from phagocytes and viruses, aids in moisture retention, and aids cell adhesion to the surface and nutrition. Gram-negative bacteria are the most prevalent carriers of the capsule. Gram-negative bacteria with capsules include *Homophiles influenza, Escherichia coli*, Klebsiella pneumonia, Salmonella, and Pseudomonas aeruginosa, Bacillus magisterium, Streptococcus pneumonia, and Streptococcus pyogenesis are samples of Gram-negative bacteria (Iddan et al., 2000).

### 3.2.3. Cell wall

The cell wall is the topmost part among many cells that shields and forms the bacterial cell. Mycoplasma is an exclusion since it does not has a cell wall. Bacterial cell walls are made up of peptidoglycan, a polysaccharide chain that has been cross-linked with unique peptides comprising D-amino acids. Bacterial cell walls are distinct from the plant as well as fungal cell walls, which are composed of cellulose and chitin, correspondingly. Bacterial cell walls are also unique than those of Archaea, as it lacks peptidoglycan (Walton, 1994).

Numerous bacteria depend on their cell wall for existence. Penicillin kills bacteria by inhibiting the cross-linking of peptidoglycan, causing the cell wall to break and lyse. Additionally, the lysozyme enzyme may destroy bacterial cell walls (Cosgrove, 2005).

Bacteria have two distinct kinds of cell walls, referred to as Gram-positive and Gram-negative. The titles derive from the way cells respond to the Gram stain, a long-used method for classifying bacterial species. Gram-positive bacteria have a rigid cell wall that is made up of several peptidoglycans and teichoic acid layers (Varner & Lin, 1989). Gram-negative bacteria, on the other hand, have a very thin cell wall composed of some layers of peptidoglycan enclosed by a second lipid membrane composed of lipopolysaccharides and lipoproteins. These structural changes may result in disparities in properties such as antibiotic susceptibility. Vancomycin, for instance, is efficient over Gram-positive bacteria but not Gram-negative pathogens like *Pseudomonas aeruginosa* or *Haemophilus influenza* (Cleland, 1971).

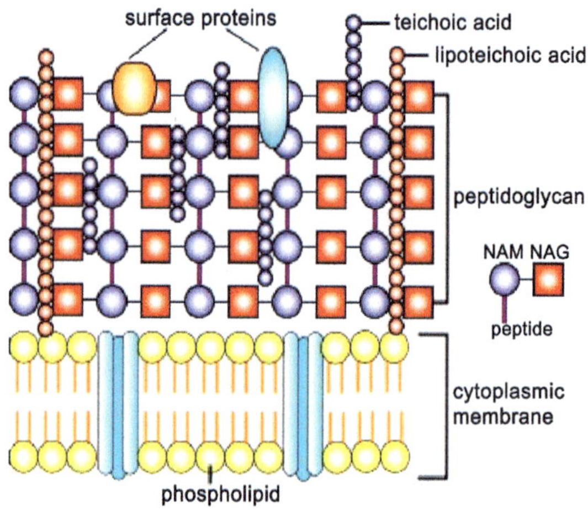


**Figure 3.2.** Gram positive bacterial cell wall.

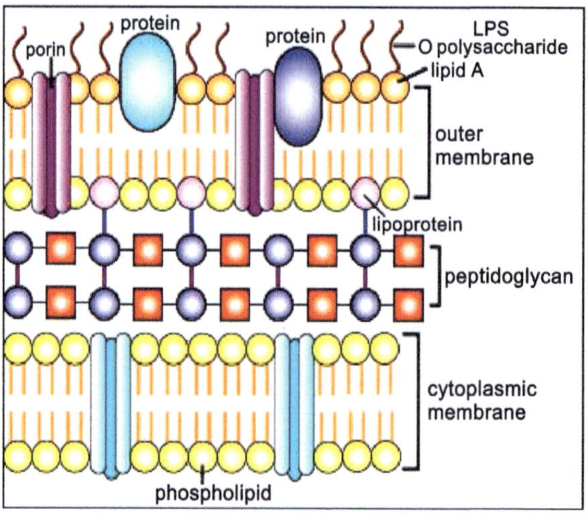


**Figure 3.3.** Gram negative bacterial cell wall.

**Source:** https://www.onlinebiologynotes.com/bacterial-cell-wall-structure-composition-types/

### 3.2.4. Cell Membrane

The cell membrane encircles the cytoplasm and controls the movement of chemicals to and from the cell. It would be covered in depth in the next sections (Yeagle, 1985).

### 3.2.5. Cytoplasm

The cytoplasm of a cell is a natural liquid that fills up the cell as well as is mostly made up of water, but it also includes enzymes, cell organelles, salts, and other organic molecules. The specifics will be detailed in the next section (Crick & Hughes, 1950).

### 3.2.6. Ribosomes

Ribosomes are the cellular organelles that are involved in protein production. Ribosomes in depth would be discussed in the next chapter (Fatica & Tollervey, 2002).

### 3.2.7. Nucleoid Region

A prokaryotic bacterium has a nucleoid area. This is the cytoplasmic region that houses the bacterial DNA molecule (Wahl et al., 2012).

### 3.2.8. Plasmids

Joshua Lederberg, an American molecular geneticist, proposed the word "plasmid" in 1952. A plasmid is DNA (found most often in bacteria) that is distinct from chromosomal DNA and could duplicate freely of it. They're round and double-stranded. Plasmids are most commonly determined in bacteria, but they can also be discovered in eukaryotic organisms. They range in size from 1 to more than 1,000 kbps (Smillie et al., 2010). The copy number counts the set of the same plasmids in individual cells, which can range from one to thousands in certain situations. Plasmids are portable since they are closely correlated with conjugation, a horizontal gene transmission mechanism. Compatible plasmids are those that could coexist inside a bacterium (Couturier et al., 1988). Plasmids that could not exist are incompatible, and they ended up lost from the cell after several centuries. Conjugative plasmids are those which encrypt one's transmission among bacteria. Such transfer genes are absent from non-conjugative plasmids, but they could be managed to carry on by conjugative plasmids through a mobilization site. They bring genetic factors responsible for a variety of metabolic actions, allowing their host bacteria to deteriorate pollutants and start producing antibacterial proteins (Carattoli, 2013).

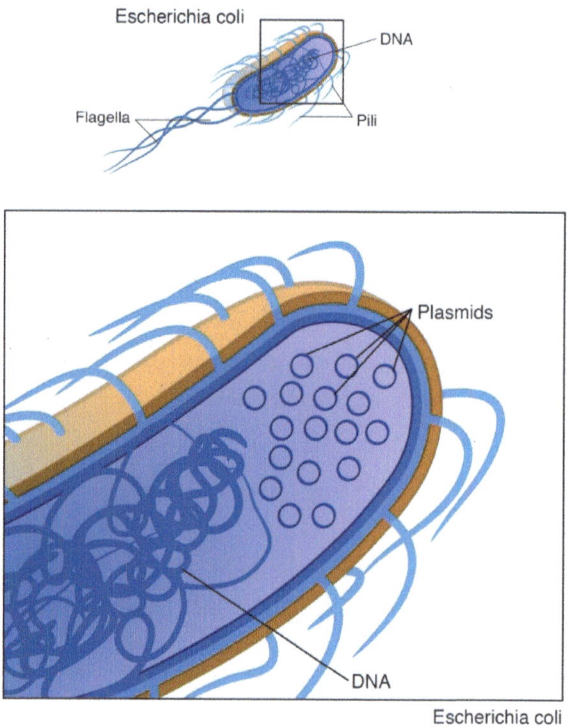


**Figure 3.4.** Graphical Representation of Plasmid.

**Source:** https://www.genome.gov/genetics-glossary/Plasmid

They can also transfer virulence genes that help in the pathogenicity of bacteria which results in diseases like plague, tetanus, anthrax, and dysentery. They are always to blame for the propagation of antibiotic resistance genes, which have a negative influence on illness treatment. The coming-up components are the different kinds of plasmids (Cohen et al., 1973).

F-plasmids for Fertility: These plasmids include tra genes and may conjugate.

Resistance (R) plasmids: They include genes that enable bacteria to generate pili and develop resistance to antibiotics or poisons.

Col plasmids are plasmids that carry genes responsible for bacteriocins, which are proteins that destroy microorganisms (Hanahan, 1983).

Degradative plasmids: Degradative plasmids allow uncommon chemicals, like toluene as well as salicylic acid, to be metabolized.

Pathogenicity plasmids: These plasmids permit the bacteria to be pathogenic.

Other kinds of plasmids include:

*Yeast integrative plasmids (YIp) are yeast vectors that survive and replicate by integrating them into the host genome.*

*Yeast Replicative Plasmid (YRp): transports a chromosomal DNA series that contains a replication origin. Because these plasmids may be destroyed at the time of budding, they are far less safe* (Barton et al., 1995).

### 3.2.9. Pili

Pili are hair-like formations on the cell membrane that aid in bacterial cell attachment. Fimbriae, which are shorter pili, aid bacteria in attaching to different surfaces. The diameter of a pilus is usually 6 to 7 nm. Conjugative pili and Type IV pili are the two kinds of pili. In the process of bacterial conjugation, conjugative pili allow DNA to be transferred among bacteria. A few pili, referred to as type IV pili, produce motile forces (Telford et al., 2006).

Pili are shorter and much more fragile than flagella. While flagella are compact, pili are much less compact and shorter as a consequence of the pressure generated by the motor. Pili may be spread across the cell's surface, or they might be clustered on one or even both sides. They're most abundant in Gram-negative bacteria, and they're present in Gram-positive bacteria and also archaea (Motzer et al., 2009).

Pili in Gram-negative bacteria may be divided into four types based on how they are assembled. Chaperone-usher pathway pili, type IV pili, curli pili, and the CS1 pilus family are the four groups. Gram-positive bacteria's pili, which as we recognize, are much fewer diverse (Haas et al., 2016). They are divided into 2 categories: narrow, slender rods and longer, more flexible pili. Streptococcus class with short and thin pili have been discovered in standard body flora, so although Corynebacterium class and pathogenic Gram-negative and Gram-positive bacteria's pili are designed inversely (Mattick, 2002). There are variances in building mechanisms among Gram-negative pili, but still, the overall trend would be that Gram-negative excretion and component assemblage takes place near the forming pilus, while Gram-positive bacteria's assembly method is separate from the pilus itself. When the pilus retracts, the pilins are often reprocessed, but when the pilus disassembles, the subunits are broken (Nowak-Sliwinska et al., 2018).

Pili can also protect bacterial cells from white blood cell threats. M-protein and pili in *Streptococcus pyogenes* may prevent phagocyte engulfment and play a part in adhesion. The F pili, which are included in the genetic exchange, are also present (Ellis et al., 2009).

Conjugation takes place in the pili. The transmission of genetic material among cells is accomplished via pili, which contains a hollow center. The fusion pilus, also known as the gender pilus or F pilus, involves receptors that let the donor's genetic data be accepted by receiving cells. In *Escherichia coli*, the F pilus may be discovered. It works by stabilizing bacteria during the conjugation process of DNA transfer (Passaniti et al., 1992).

*Pseudomonas aeruginosa* and *Myxococcus xanthus* have type IV pili, which have a variety of roles, including movement. The mobility of *P. aeruginosa* is accomplished by jerking motility and the utilization of brief periodic jerks. This form of mobility does not need the presence of many additional cells. Social gliding movement in *M. xanthus* is equivalent to twitching motion that is both based on type IV pili extension and retraction (Schmetz et al., 2002).

## 3.3. MORPHOLOGY OF PROKARYOTIC CELLS

Prokaryotic cells have few shapes; the four fundamental ones are (Gilson et al., 2003):

- Vibrio-comma-shape
- Bacilli-rod-shaped
- Cocci-spherical
- Spirochaete-spiral-shaped

### 3.3.1. Milieu

Prokaryotes are present in nearly all of the Earth's ecosystems. A few archaea and bacteria can live in harsh settings, such as great heat (thermophiles) or high salt (salinophiles) (halophiles). Extremophiles are organisms that possess these features. Many archaea are present as plankton in the oceans.

Symbiotic prokaryotes are prokaryotes that live on it or near the bodies of numerous different creatures, most notably humans (Bettelheim & Sylvester, 1948).

### 3.3.2. Sociability

Prokaryotes are thought to be exclusively single cells, yet they may make stable organized colonies termed "biofilms" in a stabilizing polymer matrix. In place and time, cells in biofilms typically display diverse gene expression patterns designs (phenotypic differentiation) (Simmel, 1949). These variations in production occur as a consequence of quorum sensing or cell-to-cell signaling transmission, much as they do in multicellular eukaryotes. Microcolonies are microbial biofilms that are composed of roughly dome-shaped aggregates of bacteria and matrix divided by "voids" whereby the environment (water) might pass fairly unhindered. The microcolonies can create a closed layer just above the substratum, shutting the system of pathways that separate microcolonies (Weiss, 1969). Bacterial biofilms may also be 100 times highly antibiotic-resistant as compared to the free-living single cells and could be harder to eliminate from the surface once colonized. Additional features of bacterial collaboration, such as bacterial conjugation and quorum-sensing-mediated pathogenicity, give further problems for researchers and medical practitioners attempting to address the illnesses they cause (Schiffrin, 1984).

### 3.3.3. Colony of Bacteria

The majority of bacteria exist in colonies. Living things of similar species that live close in a mutualistic relationship are referred to as a colony. Genetically, all organisms in a group are identical. The colony's form might be round or irregular. Bacterial colonies usually seem glossy and smooth. Colony-forming unit (CFU) is a microbiology term that refers to the number of live bacteria in these kinds of colonies. If a bacterial cell such as *Escherichia coli* splits per 20 minutes, a colony will have 230 or 1048576 cells by 30 cell divisions (Pinhassi et al., 1997).

### 3.3.4. Reproduction

Binary fission is an asexual reproductive way used by bacteria and archaea. Binary fission is an asexual reproductive way. The genomic DNA is replicated at the time of binary fission, and the parent cell is split into two homologous chromosomes. All creatures in a colony are biologically equal because of binary fission (Figure below) (Bourdieu, 2018). DNA replication is proceeded by DNA partition, partition site selection, cellular membrane invagination, and creation of a fresh cell wall, all of which have been closely regulated by cellular proteins. The protein FtsZ, which assembles into a ring-like structure in the cell's core, is a crucial part of cellular division. The FtsZ ring is where the rest of the divisional equipment links together. This machinery is placed as division separates the cytoplasm at the time of evading damage to DNA. The cytoplasm is cleft in two at the time of division, and the latest cell wall is formed (Johnson, 2015).

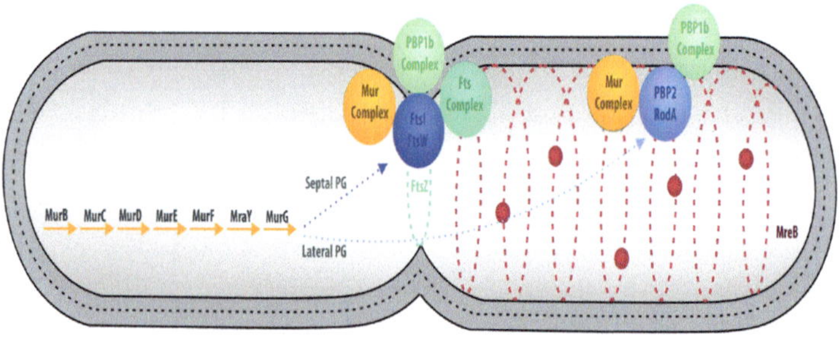


**Figure 3.5.** Binary fission in prokaryotes.

**Source:** Rivaș-Marín E, Canosa I and Devos DP (2016) Evolutionary Cell Biology of Division Mode in the Bacterial Planctomycetes-Verrucomicrobia-Chlamydiae *Superphylum.* Front. Microbiol. *7:1964.* doi: 10.3389/fmicb.2016.01964

## 3.3.5. Products/Application

Prokaryotes contribute to the production of yogurt, cheese, sour cream, and antibiotics, among other things. They serve as a repository for a variety of industrially significant enzymes, including lipases, proteases, and amylases, that are used in detergents, paper, as well as leather companies (Earnest & Corsi, 2013).

## 3.4. EUKARYOTE

A eukaryotic cell is made up of membrane-bound organelles. Eukaryota is their classification. Mammalian, plant, and fungal eukaryotes, and the bulk of protist microorganisms, are enormously complex organisms. Monoecious eukaryotes appear to be monoecious between the three phases of life (organisms that form a clade). Bacteria and Archaea, some other two kinds of prokaryotes, have neither of the properties listed above (Martin & Müller, 1998; Green et al., 2004). Eukaryotes denote a small percentage of all life forms; in reality, bacteria are 10 times more abundant in a person than cells. Despite this, because of their substantially larger size, their global biomass is thought to be equivalent to that of prokaryotes. In contrast to prokaryotes, eukaryotes' genome is housed within the nucleus and guarded by a nuclear membrane. Many membrane-bound organelles dwell in the cytoplasmic of their cells, in addition to the nuclei. In cellular division, the cytoskeleton coordinates movements that split the genome and is in the form of tightly packed compacted structures called chromosomes (Cavalier-Smith, 1981).

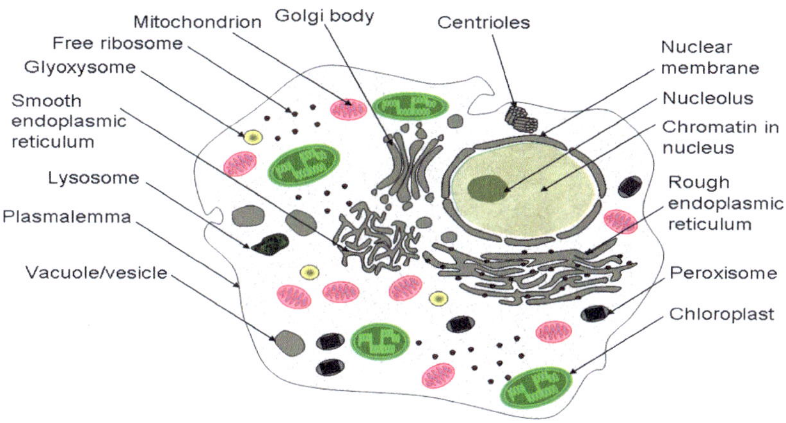


**Figure 3.6.** Structure of eukaryotic cell.

**Source:** https://www.nagwa.com/en/explainers/297140534241/3.

### 3.4.1. Classification of Eukaryotic Cell

Four domains comprise the eukaryotes (Singh et al., 2020):

- Kingdom Protista
- Kingdom Plantae
- Kingdom Animalia
- Kingdom Fungi

### 3.4.2. Cell Features

The size of eukaryotic cells is significantly greater compared to that of prokaryotic cells. About ten and one hundred micrometers. They include a range of different membranes and structures known as organelles, as well as a cytoskeleton formed of microtubules, microfilaments, and intermediate filaments that contribute significantly to the organization and form of the cell. In nuclear division, eukaryotic DNA is partitioned into numerous linear strands called chromosomes that are segregated by a microtubular spindle (Ziegler & Groscurth, 2004).

### 3.4.3. Internal Membrane

Eukaryote cells have some membrane-bound components commonly known as the endomembrane system, which perform a range of activities. By splitting off other walls, easy compartments known as vesicles or vacuoles may develop. Several cells consume food and other things by endocytosis, a procedure in which the outer sheath corresponds and clamps off to create a vesicle. Almost all other membrane-bound structures are most likely descended from similar vesicles (Veillette et al., 1988). A dual membrane (often known as a nuclear envelope) surrounds the nucleus, containing holes that enable the material to pass into and out. The endoplasmic reticulum, or ER, is formed by different tube and sheet-like expansions of the nuclear membrane but also is included in protein transport and maturation. It contains the rough ER, wherein ribosomes are anchored to produce proteins that reach the lumen. Following that, they usually enter vesicles, which sprout from the smooth ER (Bamford et al., 1995). Such protein-carrying vesicles are discharged and subsequently changed in stacks of flattened vesicles known as Golgi bodies or dictyosomes for most eukaryotes. Vesicles may be tailored to serve a variety of functions. Lysosomes, for example, include enzymes that destroy the material of food vacuoles, while peroxisomes are being used to dissolve peroxide, which would otherwise be harmful (Eimhjellen et al., 1967). Many protozoa feature contractile vacuoles that gather and release surplus water, as well as extrusomes that eject substances needed to ward off predators or catch food. Hormones are often generated in vesicles in complex organisms. In-plant species, a central vacuole takes up the majority of a cell's volume and is principally responsible for maintaining the cell's osmotic pressure. In the next lessons, the different cell organelles would be described in depth (Strömsten et al., 2005).

### 3.4.4. Reproduction

Cell division and nuclear division are frequently synchronized. Mitosis, a procedure wherein every daughter nucleus receives 1 set of chromosomes, is usually used to accomplish this. For most eukaryotes, sexual reproduction occurs by nuclear fusion (syngamy) and meiosis, which normally involves an alternating among haploid generations, in which only a single duplicate of every chromosome is available, and diploid generations, in which two are available. This design has a lot of diversity in it (McNeilly, 2001).

### 3.4.5. Association/Hierarchy

Cells interact to create tissue as in animal as well as plant kingdoms, and tissue develops into organs, that eventually constitute the entire creature (Coleman, 1963).

## 3.5. PROKARYOTES VERSUS EUKARYOTES

Here is a list of the differences between prokaryotes and eukaryotes. Eukaryotes have such a reduced metabolic efficiency as well as a greater reproduction rate as compared to the prokaryotes due to their reduced surface area to volume ratio. Cells specialized for metabolism, like intestinal vili, get a larger surface area in certain multicellular species (Kozak, 1999).

**Table 3.1.** Difference between prokaryotes and eukaryotes

| Characteristic | Prokaryotes | Eukaryotes |
|---|---|---|
| Nucleus | Not any nuclear membrane or nucleoli(nucleoid) | Exact nucleus, containing nuclear membrane & nucleoli |
| Sizeofcell | Generally 0.2-2.0 mm in \s diameter | Generally 10-100 mm in diameter |
| Cell wall | Frequently express; chemically complicated (typical bacterial cell wall involves peptidoglycan) | Catalytically easy when existing |
| Flagella | It is made up of two protein-building units. | Complex; include a large number of microtubules |
| Cytoplasm | Nocytoskeleto-norcytoplasmic streaming | Cytoskeleton; cytoplasmic streaming |
| Plasma membrane | No carbohydrates as well as typically without Sterols | Sterols and carbohydrates which act as receptors present |

| Ribosomes | The smaller diameter. | Larger size(80S); smaller size(70S) in organelles |
|---|---|---|
| Sexual reproduction | Nomeiosis; transmission of DNA<br>fragments (conjugation) | Includes Meiosis |
| Cell division | Binary fission | Mitosis |

### 3.5.1. Phytoplanktons and Zooplanktons

Phytoplankton is tiny photosynthesizing creatures that live in the topmost sunlight level of practically all seas and bodies of freshwater resources as well as receive their energy via photosynthesis. Amazingly, phytoplankton is responsible for almost partial of any photosynthetic impact on Earth (Mullin & Brooks, 1976). Phytoplankton includes bacteria, protists, and the majority of single-celled plants. Cyanobacteria, silica-encased diatoms, dinoflagellates, green algae, and chalk-coated coccolithophores are some of the more frequent types (Irigoien et al., 2004). The development of phytoplankton is dependent on the presence of carbon dioxide, sunshine, and nutrients. Phytoplankton needs varying amounts of nutrients like nitrate, phosphate, silicate, and calcium. Certain phytoplankton species are nitrogen fixers and may flourish in environments with low nitrate concentrations. Additionally, they need small levels of iron, that restrict phytoplankton development in wide regions of the ocean due to reduced iron concentrations (Jakhar, 2013).

Zooplankton is a term that refers to a diverse mix of tiny protozoans and giant metazoans. It encompasses both holoplanktonic species that spent their whole lifetimes in the plankton and meroplanktonic creatures that pass a portion of their lifespan in the plankton before actually moving to nekton or a sessile, benthic lifestyle. While zooplankton is predominantly moved by ambient water currents, some species possess mobility, which they exploit to evade predators (just like in diel vertical migration) and to improve the rate of prey contact (Dokulil & Herzig, 2009).

### 3.5.2. Comparing Prokaryotes and Eukaryotes

All species on Earth are composed of either prokaryotic or eukaryotic cells. Prokaryotes would be the first living organisms. Eukaryotes are supposed to have been created from prokaryotes a few 2.7 billion - years - old, concerning scientists (Kaul et al., 2004).

The key dissimilarity between these two types of organisms is that eukaryotic cells have a membrane-bound nucleus while prokaryotic cells have not. Eukaryotes store genetic data in their nuclei. DNA is tightly compacted collectively in prokaryotes' nucleoid region, but it is not contained within a membrane-bound nucleus (Mirande, 1991).

The nucleus is only one of multiple membrane-bound organelles present in eukaryotes. Prokaryotes, on the other hand, lacking membrane-bound organelles. The DNA sequence is the other key difference. Eukaryotic DNA is composed of many double-stranded linear DNA particles, while prokaryotic DNA is composed of double-stranded circular DNA particles (Nichols et al., 2004).

### 3.5.3. Key Similarities between Prokaryotes and Eukaryotes

All cells, whether prokaryotic or eukaryotic, share these four features (Kisselev, 2002):

1. DNA
2. Plasma membrane
3. Cytoplasm
4. Ribosomes

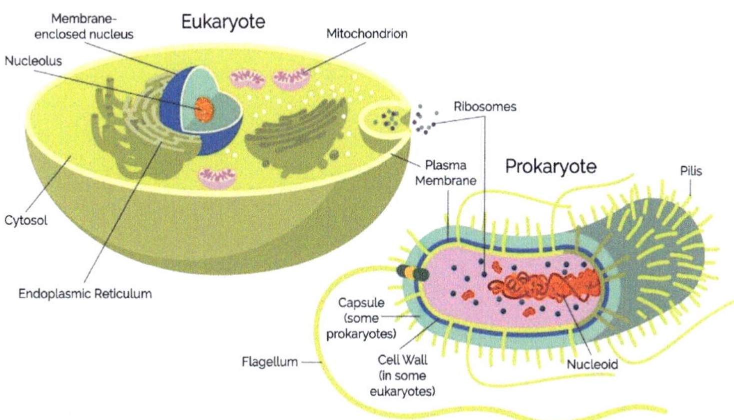


**Figure 3.7.** Comparison showing the shared and unique features of prokaryotes and eukaryotes.

**Source:** https://www.technologynetworks.com/cell-science/articles/prokaryotes-vs-eukaryotes-what-are-the-key-differences-336095

### 3.5.4. Transcription and Translation in Prokaryotes vs Eukaryotes

Translation and transcription are tightly connected in prokaryotic organisms, which means that translation starts along with mRNA production (Roussel & Zhu, 2006).

Transcription and translation are not covalently connected in eukaryotic cells. Transcription happens in the nucleus and output in the formation of mRNA. Afterward, the mRNA leaves the nucleus as well as is translated into the cytoplasm of the cell (French et al., 2007).

### 3.5.5. Key Differences between Prokaryotes and Eukaryotes

Prokaryotes and eukaryotes vary in many methods, involving morphological differences like either cell does have a nucleus or either it contains membrane-bound organelles, and molecular variations like either the DNA is round or straight. The distinctions are shown in the list below (Ren & Paulsen, 2005).

**Table 3.2.** A detailed comparison between Prokaryotes and Eukaryotes

| S.N. | Character | Prokaryotes | Eukaryotes |
|---|---|---|---|
| | Cell Type | Generally having single cell (few cyanobacteria might be multicellular) | Usually multicellular |
| | Nucleus Location | Able in the cytoplasm, close to mesosomes | present in the membrane-bound structure |
| | Major groups | Bluegreen Algae, Bacteria, and Archie | Fungi, animals, protozoa, plants, algae, |
| | Genes | Represented in clusters known as operons. | Presented separately. |
| | Size (approximate) | 0.5 to 3.0 μm | >5 μm |
| | Term Origin | Greek for “primitive nucleus” | Greek for “true nucleus” |
| | Complexity | Simple | Complex organization. |

| | | | |
|---|---|---|---|
| | Definition | Creatures are composed of the cell(s) that does not a cell nucleus or any membrane-encased organelles. | Organisms are composed of cells and contain a membrane-bound nucleus and membrane-bound organelles. |
| | Nucleolus | Absent | Present |
| | Nuclear membrane | No nuclear membrane. | Typical membrane present. |
| | Chromosome shape | Circular | Linear |
| | Origin | Around 3.5 billion years ago. | Around 2 billion years ago |
| | Genome nature | Effective and solid with slight repetitive DNA | With great volumes of non-coding repetitive DNA. |
| | DNA base ratio (G+C %) | 28 to 73 | almost 40 |
| | Genome | .DNA haploid genome | DNA diploid genome |
| | Cytoskeleton | May be absent | Present |
| | Ribosome's location | Able in the cytoplasm or bound to the cell membrane | Close to the uneven endoplasmic reticulum |
| | Cytoplasmic streaming | Not present | Available |
| | Golgi bodies | Not present | Available |
| | Endoplasmic reticulum | Not present | Available |
| | Mesosomes | Present. Handles the functions of Golgi bodies and mitochondria, as well as assisting in chromosomal segregation in cell division. | Not present |
| | Centrosome | Not present | Present excluding flowering plants. |

| | | | |
|---|---|---|---|
| | Microtubules | Absent or rare | Present |
| | Chloroplasts | Not present; chlorophyll distributed in the cytoplasm | Available (in plants) |
| | Peroxisomes | Not present | Available |
| | Ribosomes (sedimentation coefficient) | 70S (50S + 30S). Smaller. | 80S (60S + 40S). Larger. |
| | Glycocalyx | Available | A few |
| | Mitochondria | Not present | Available |
| | Cytoplasmic membrane | Does not have sterols (except *Mycoplasma*) | Contain sterols |
| | Cell wall | Complex structure containing protein, lipids, and peptidoglycans | Present for plant cells and fungi; otherwise absent |
| | Muramic acid | Present | Absent |
| | Movement | Simple flagellum, if present | Complex flagellum, if present |
| | Respiration | Via cytoplasmic membrane | Via mitochondria |
| | Metabolic rate | Higher due to larger surface area to volume ratio | Comparatively slow |
| | Reproduction | Asexual (binary fission) | Sexual and asexual/ Mitotic division |
| | Genetic Recombination | Partial, unidirectional transfer | Meiosis and fusion of gametes |
| | Zygote | Merozygotic (partially diploid) | Diploid |
| | Extrachromosomal DNA | Plasmid | Inside the mitochondria |
| | DNA replication | Occurs in the cytoplasm. | Occurs in the nucleus. |

## REFERENCES


1. Apirion, D., & Miczak, A. (1993). RNA processing in prokaryotic cells. Bioessays, 15(2), 113-120.
2. Bamford, D. H., Caldentey, J., & Bamford, J. K. (1995). Bacteriophage PRD1: a broad host range dsDNA tectivirus with an internal membrane. Advances in virus research, 45 (1), 281-319.
3. Barton, B. M., Harding, G. P., & Zuccarelli, A. J. (1995). A general method for detecting and sizing large plasmids. Analytical Biochemistry, 226(2), 235-240.
4. Berg, H. C. (2003). The rotary motor of bacterial flagella. Annual review of biochemistry, 72(1), 19-54.
5. Bettelheim, B., & Sylvester, E. (1948). A therapeutic milieu. American Journal of Orthopsychiatry, 18(2), 191.
6. Boiteux, S., Coste, F., & Castaing, B. (2017). Repair of 8-oxo-7, 8-dihydroguanine in prokaryotic and eukaryotic cells: Properties and biological roles of the Fpg and OGG1 DNA N-glycosylases. Free Radical Biology and Medicine, 107(1), 179-201.
7. Bourdieu, P. (2018). Cultural reproduction and social reproduction. In Knowledge, education, and cultural change (Vol. 1, pp. 71-112). Routledge.
8. Carattoli, A. (2013). Plasmids and the spread of resistance. International journal of medical microbiology, 303(6-7), 298-304.
9. Cavalier-Smith, T. (1975). The origin of nuclei and of eukaryotic cells. Nature, 256(5517), 463-468.
10. Cavalier-Smith, T. (1981). Eukaryote kingdoms: seven or nine?. Biosystems, 14(3-4), 461-481.
11. Cleland, R. (1971). Cell wall extension. Annual review of plant physiology, 22(1), 197-222.
12. Cohen, S. N., Chang, A. C., Boyer, H. W., & Helling, R. B. (1973). Construction of biologically functional bacterial plasmids in vitro. Proceedings of the National Academy of Sciences, 70(11), 3240-3244.
13. Coleman, E. B. (1963). The association hierarchy as an indicator of extraexperimental interference. Journal of Verbal Learning and Verbal Behavior, 2(5-6), 417-421.
14. Cosgrove, D. J. (2005). Growth of the plant cell wall. Nature reviews Molecular cell biology, 6(11), 850-861.

15. Couturier, M. A. R. T. I. N. E., Bex, F., Bergquist, P. L., & Maas, W. K. (1988). Identification and classification of bacterial plasmids. Microbiological Reviews, 52(3), 375-395.
16. Crick, F. H. C., & Hughes, A. F. W. (1950). The physical properties of cytoplasm. Experimental cell research, 1(1), 37-80.
17. Dokulil, M. T., & Herzig, A. (2009). An analysis of long-term winter data on phytoplankton and zooplankton in Neusiedler See, a shallow temperate lake, Austria. Aquatic Ecology, 43(3), 715-725.
18. Earnest, C. M., & Corsi, R. L. (2013). Inhalation exposure to cleaning products: application of a two-zone model. Journal of occupational and environmental hygiene, 10(6), 328-335.
19. Eimhjellen, K. E., Steensland, H., & Traetteberg, J. (1967). A *Thiococcus* sp. nov. gen., its pigments and internal membrane system. Archiv für Mikrobiologie, 59(1), 82-92.
20. Ellis, L., Hammers, H., & Pili, R. (2009). Targeting tumor angiogenesis with histone deacetylase inhibitors. Cancer letters, 280(2), 145-153.
21. Fatica, A., & Tollervey, D. (2002). Making ribosomes. Current opinion in cell biology, 14(3), 313-318.
22. French, S. L., Santangelo, T. J., Beyer, A. L., & Reeve, J. N. (2007). Transcription and translation are coupled in Archaea. Molecular biology and evolution, 24(4), 893-895.
23. Fröhlich, J., & König, H. (2000). New techniques for isolation of single prokaryotic cells. FEMS microbiology reviews, 24(5), 567-572.
24. Gibbons, I. R. (1981). Cilia and flagella of eukaryotes. The Journal of cell biology, 91(3), 107s-124s.
25. Gilson, P. R., Yu, X. C., Hereld, D., Barth, C., Savage, A., Kiefel, B. R., ... & Beech, P. L. (2003). Two Dictyostelium orthologs of the prokaryotic cell division protein FtsZ localize to mitochondria and are required for the maintenance of normal mitochondrial morphology. Eukaryotic Cell, 2(6), 1315-1326.
26. Green, J. L., Holmes, A. J., Westoby, M., Oliver, I., Briscoe, D., Dangerfield, M., ... & Beattie, A. J. (2004). Spatial scaling of microbial eukaryote diversity. Nature, 432(7018), 747-750.
27. Haas, N. B., Manola, J., Uzzo, R. G., Flaherty, K. T., Wood, C. G., Kane, C., ... & DiPaola, R. S. (2016). Adjuvant sunitinib or sorafenib for high-risk, non-metastatic renal-cell carcinoma (ECOG-ACRIN

E2805): a double-blind, placebo-controlled, randomised, phase 3 trial. The Lancet, 387(10032), 2008-2016.

28. Hanahan, D. (1983). Studies on transformation of Escherichia coli with plasmids. Journal of molecular biology, 166(4), 557-580.
29. Iddan, G., Meron, G., Glukhovsky, A., & Swain, P. (2000). Wireless capsule endoscopy. Nature, 405(6785), 417-417.
30. Irigoien, X., Huisman, J., & Harris, R. P. (2004). Global biodiversity patterns of marine phytoplankton and zooplankton. Nature, 429(6994), 863-867.
31. Jakhar, P. (2013). Role of phytoplankton and zooplankton as health indicators of aquatic ecosystem: A review. International Journal of Innovation Research Study, 2(12), 489-500.
32. Johnson, A. L. (2015). Reproduction in the female. In Sturkie's avian physiology (Vol. 1, pp. 635-665). Academic Press.
33. Kaul, M., Barbieri, C. M., & Pilch, D. S. (2004). Fluorescence-based approach for detecting and characterizing antibiotic-induced conformational changes in ribosomal RNA: comparing aminoglycoside binding to prokaryotic and eukaryotic ribosomal RNA sequences. Journal of the American Chemical Society, 126(11), 3447-3453.
34. Kisselev, L. (2002). Polypeptide release factors in prokaryotes and eukaryotes: same function, different structure. Structure, 10(1), 8-9.
35. Kozak, M. (1999). Initiation of translation in prokaryotes and eukaryotes. Gene, 234(2), 187-208.
36. Krämer, R. (1996). Analysis and modeling of substrate uptake and product release by prokaryotic and eukaryotic cells. Metabolic Engineering, Vol. 1, pp. 31-74.
37. Machin, K. E. (1958). Wave propagation along flagella. Journal of Experimental Biology, 35(4), 796-806.
38. Macnab, R. M. (2003). How bacteria assemble flagella. Annual Reviews in Microbiology, 57(1), 77-100.
39. Martin, W., & Müller, M. (1998). The hydrogen hypothesis for the first eukaryote. Nature, 392(6671), 37-41.
40. Mattick, J. S. (2002). Type IV pili and twitching motility. Annual Reviews in Microbiology, 56(1), 289-314.
41. McNeilly, A. S. (2001). Lactational control of reproduction. Reproduction, Fertility and Development, 13(8), 583-590.

42. Mirande, M. (1991). Aminoacyl-tRNA synthetase family from prokaryotes and eukaryotes: structural domains and their implications. Progress in nucleic acid research and molecular biology, 40(1), 95-142.
43. Moens, S., & Vanderleyden, J. (1996). Functions of bacterial flagella. Critical reviews in microbiology, 22(2), 67-100.
44. Motzer, R. J., Hutson, T. E., Tomczak, P., Michaelson, M. D., Bukowski, R. M., Oudard, S., ... & Figlin, R. A. (2009). Overall survival and updated results for sunitinib compared with interferon alfa in patients with metastatic renal cell carcinoma. Journal of clinical oncology, 27(22), 3584.
45. Mullin, M. M., & Brooks, E. R. (1976). Some consequences of distributional heterogeneity of phytoplankton and zooplankton 1. Limnology and Oceanography, 21(6), 784-796.
46. Nayak, D. P., Davis, A. R., McQueen, N. L., Bos, T. J., Jabbar, M. A., Sivasubramanian, N., & Lionelli, G. (1985). Biological and immunological properties of haemagglutinin and neuraminidase expressed from cloned cDNAs in prokaryotic and eukaryotic cells. Vaccine, 3(3), 165-171.
47. Nichols, C. E., Ren, J., Leslie, K., Dhaliwal, B., Lockyer, M., Charles, I., ... & Stammers, D. K. (2004). Comparison of ligand-induced conformational changes and domain closure mechanisms, between prokaryotic and eukaryotic dehydroquinate synthases. Journal of molecular biology, 343(3), 533-546.
48. Nowak-Sliwinska, P., Alitalo, K., Allen, E., Anisimov, A., Aplin, A. C., Auerbach, R., ... & Griffioen, A. W. (2018). Consensus guidelines for the use and interpretation of angiogenesis assays. Angiogenesis, 21(3), 425-532.
49. Oikonomou, C. M., Chang, Y. W., & Jensen, G. J. (2016). A new view into prokaryotic cell biology from electron cryotomography. Nature Reviews Microbiology, 14(4), 205-220.
50. Passaniti, A., Taylor, R. M., Pili, R., Guo, Y., Long, P. V., Haney, J. A., ... & Martin, G. R. (1992). A simple, quantitative method for assessing angiogenesis and antiangiogenic agents using reconstituted basement membrane, heparin, and fibroblast growth factor. Laboratory investigation; a journal of technical methods and pathology, 67(4), 519-528.
51. Pinhassi, J., Zweifel, U. L., & Hagstroëm, A. (1997). Dominant marine

bacterioplankton species found among colony-forming bacteria. Applied and Environmental Microbiology, 63(9), 3359-3366.

52. Ren, Q., & Paulsen, I. T. (2005). Comparative analyses of fundamental differences in membrane transport capabilities in prokaryotes and eukaryotes. PLoS computational biology, 1(3), pp. 27.
53. Roussel, M. R., & Zhu, R. (2006). Validation of an algorithm for delay stochastic simulation of transcription and translation in prokaryotic gene expression. Physical biology, 3(4), 274.
54. Sacchetti, A., Cappetti, V., Marra, P., Dell'Arciprete, R., El Sewedy, T., Crescenzi, C., & Alberti, S. (2001). Green fluorescent protein variants fold differentially in prokaryotic and eukaryotic cells. Journal of Cellular Biochemistry, 81(S36), 117-128.
55. Schiffrin, D. (1984). Jewish argument as sociability. Language in society, 13(3), 311-335.
56. Schippers, A., Neretin, L. N., Kallmeyer, J., Ferdelman, T. G., Cragg, B. A., John Parkes, R., & Jørgensen, B. B. (2005). Prokaryotic cells of the deep sub-seafloor biosphere identified as living bacteria. Nature, 433(7028), 861-864.
57. Schmetz, J., Pili, P., Tjemkes, S., Just, D., Kerkmann, J., Rota, S., & Ratier, A. (2002). An introduction to Meteosat second generation (MSG). Bulletin of the American Meteorological Society, 83(7), 977-992.
58. Silverman, M., & Simon, M. I. (1977). Bacterial flagella. Annual review of microbiology, 31(1), 397-419.
59. Simmel, G. (1949). The sociology of sociability. American journal of sociology, 55(3), 254-261.
60. Singh, S., Rathva, H. K., Sahay, T., Dhanjal, D. S., Chopra, C., & Singh, R. (2020). *Gemmata obscuriglobus*: A connecting link between prokaryotic and eukaryotic cell. Biologia, 75(12), 2433-2439.
61. Smillie, C., Garcillán-Barcia, M. P., Francia, M. V., Rocha, E. P., & de la Cruz, F. (2010). Mobility of plasmids. Microbiology and Molecular Biology Reviews, 74(3), 434-452.
62. Strömsten, N. J., Bamford, D. H., & Bamford, J. K. (2005). In vitro DNA packaging of PRD1: a common mechanism for internal-membrane viruses. Journal of molecular biology, 348(3), 617-629.
63. Telford, J. L., Barocchi, M. A., Margarit, I., Rappuoli, R., & Grandi, G.

(2006). Pili in gram-positive pathogens. Nature Reviews Microbiology, 4(7), 509-519.

64. Varner, J. E., & Lin, L. S. (1989). Plant cell wall architecture. Cell, 56(2), 231-239.
65. Veillette, A., Bookman, M. A., Horak, E. M., & Bolen, J. B. (1988). The CD4 and CD8 T cell surface antigens are associated with the internal membrane tyrosine-protein kinase p56lck. Cell, 55(2), 301-308.
66. Wahl, A., Schuth, N., Pfeiffer, D., Nussberger, S., & Jendrossek, D. (2012). PHB granules are attached to the nucleoid via PhaM in *Ralstonia eutropha*. BMC microbiology, 12(1), 1-11.
67. Walton, J. D. (1994). Deconstructing the cell wall. Plant physiology, 104(4), 1113.
68. Weiss, R. S. (1969). The fund of sociability. Trans-action, 6(9), 36-43.
69. Yeagle, P. L. (1985). Cholesterol and the cell membrane. Biochimica et Biophysica Acta (BBA)-Reviews on Biomembranes, 822(3-4), 267-287.
70. Ziegler, U., & Groscurth, P. (2004). Morphological features of cell death. Physiology, 19(3), 124-128.

CHAPTER 4

# PROPERTIES AND FUNCTIONS OF PROTEINS

CONTENTS

# 4.1. INTRODUCTION

Protein, a complex molecule found in all living things, is found in all living things. Proteins are nutrient-dense and serve a critical role in the chemical reactions that keep life afloat. Proteins were identified as important by early nineteenth-century chemists, notably Swedish scientist Jöns Jacob Berzelius, who coined the term protein in 1838, a word derived from the Greek parties, which means "first place." Proteins differ from one species to the next because they are species-specific. They are also organ-specific; muscle proteins differ from brain and liver proteins within the same organism.

A protein molecule is far larger than a sugar or salt molecule, and it is made up of a lengthy chain of amino acids connected like beads on a string. Proteins are made up of about twenty different amino acids that occur naturally. Amino acid composition and order have been equivalent in proteins with alike activities.

Even though the characteristics of the amino acids that contain proteins may be recognized to establish structural-functional correlations, it is currently not possible to clarify all of a protein's activities from its amino acid order (Isralewitz et al., 2001; Hu et al., 2011).

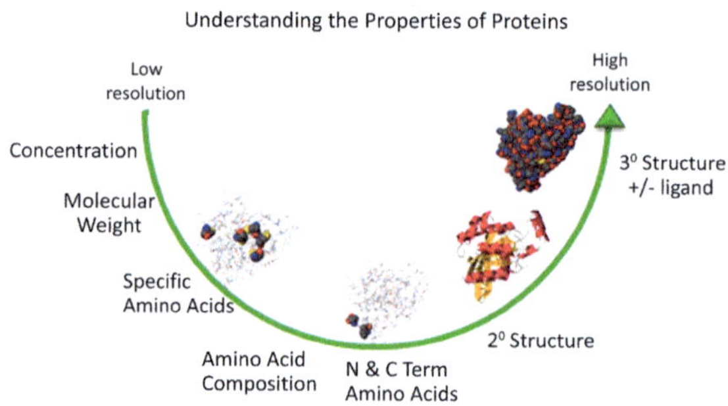


**Figure 4.1.** Protein Characteristics.

**Source:** https://employees.csbsju.edu/hjakubowski/classes/ch331/protstructure/olcompseqconform.html

Animal organs often have a significantly greater protein concentration than blood plasma. Muscles, for instance, comprise roughly thirty percent of protein, whereas the liver has 20% to 30% and red blood cells comprise 30%. Bones, hair, and other tissues and organs having lower water composition contain higher protein percentages. In mammals, the number of peptides and free amino acids is substantially lower than the protein quantity;

protein molecules have been made in cells via step to step arrangement of amino acids and discharged into the bodily fluids only once biosynthesis has been completed (Roitberg et al., 1995; Gosline et al., 2002).

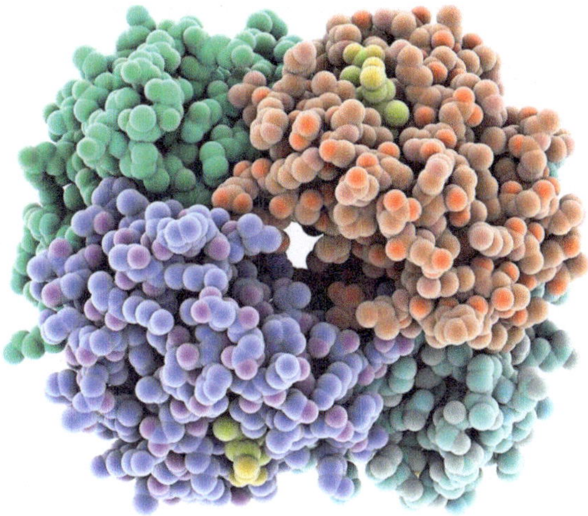

**Figure 4.2.** The Cell's Proteins.

**Source:** https://www.thoughtco.com/protein-function-373550

The high protein concentration of specific organs doesn't imply that protein significance is proportional to their quantity in a tissue or organism; on the contrary, among the most specific proteins, like hormones and enzymes, are found in minute quantities. Proteins' significance stems mostly from their function (Cohn, 1941). So yet, all enzymes discovered have been proteins. All metabolic processes are catalyzed by enzymes, allowing a living being to make up the chemical compounds required for life, like proteins, lipids, carbohydrates, and nucleic acids as well as change and break down them. It is impossible to live without enzymes. Several protein hormones have key regulatory roles. The respiratory protein haemoglobin serves as an oxygen carrier in the blood throughout all vertebrates and transports oxygen from the lungs to the body's tissues and organs. The structure of the animal body is maintained and protected by a broad set of structural proteins (Zoraghi et al., 2004; Schwartz et al., 2015).

## 4.2. CLASSIFICATION OF PROTEINS

Since the protein's biological function was not yet developed, an effort had been made to categorize proteins based on their physical and chemical characteristics after 2 German chemists, Emil Fischer and Franz Hofmeister, autonomously mentioned in 1902 that proteins have been primarily polypeptides comprising several amino acids. (It wasn't until the 1920s that the protein nature of enzymes had been established.) Proteins had been generally categorized based on their solubility in a variety of solvents. Furthermore, such categorization is not valid since proteins with very distinct functions and structures can have comparable solubilities, Proteins with comparable structures and activities, on the other hand, might have vastly different solubilities. Meanwhile, the words connected with the previous categorization are still extensively utilized (Habermann, 2004).

Albumins are water-soluble proteins and ammonium sulfate-saturated water. Globulins, on either side, are salted out (that is precipitated) when ammonium sulphate is half-saturated. Pseuglobulins are globulins that have been soluble in salt-free water while euglobulins are globulins that have not been soluble in freshwater. Plant proteins glutelins and prolamins are not soluble in water; prolamins disintegrate in fifty to eighty percent ethanol, glutelins inside an alkaline or acidified solution. The word protamine refers to a group of proteins found in fish sperm which contain around 80% arginine and are hence very alkaline (Sarkar & Sen, 2004). Histones, that are less alkaline, are thought to exclusively exist in cell nuclei and are attached to nucleic acids.

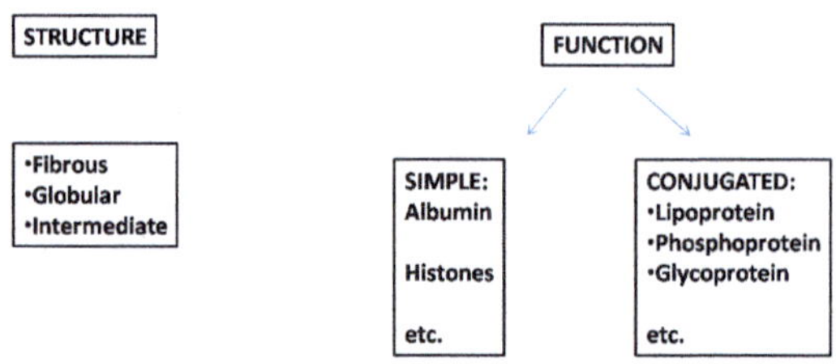


**Figure 4.3.** Protein Categorization.

**Source:** Self-made

The insoluble proteins of animal organs are referred to as scleroproteins. Keratin, an insoluble protein found in epithelial tissues including hair and skin, and collagen, a connective tissue protein, are two among them. Although conjugated proteins are complicated protein molecules with both nonprotein and protein components, they have been referred to as conjugated proteins. The prosthetic group is the nonprotein component. Mucoproteins, comprise carbohydrates and protein; lipoproteins, which comprise lipids; phosphoproteins, which have been high in phosphate; chromoproteins, that comprise pigments like carotenoids, melanin, iron-porphyrins, and bile pigments; and ultimately, nucleoproteins, that comprise a nucleic acid (Murzin, 1996; Vestal & Jeyaratnam, 2011).

## 4.3. PROTEIN GENERAL STRUCTURE AND CHARACTERISTICS

### 4.3.1. Structure of Animo Acids

The majority of proteins are composed of longer chains of α-amino acids, which is a characteristic of most proteins. The α-amino acids receive their name from the fact that the molecule's α-carbon atom has both an amino group (—$NH_2$) (—COOH) and a carboxyl group (—COOH), which are both present on the molecule (Lo Conte et al., 2000).

In acidic solutions, the —COO groups interact with hydrogen ions (H+) and are changed into their uncharged form whenever pH goes below the 4 carboxyl group. Ammonium groups (—$NH^{+3}$) lose a hydrogen ion and are transformed into amino groups in basic solutions with a pH of more than nine. since amino acids in the pH range of 4 to 8 have both a +ve and -ve charge, they don't move when subjected to an electrical field. Dipolar ions, also called zwitterions, are a form of creation that fits into this category (such as hybrid ions) (Murzin et al., 1995; Hubbard et al., 1997 ).

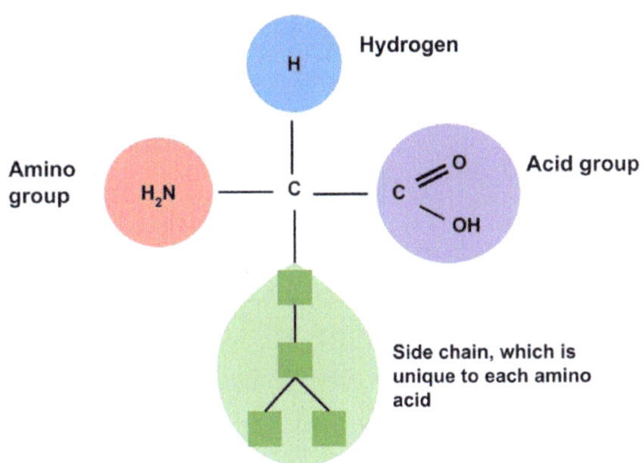


**Figure 4.4.** The Structure of amino Acid.

**Source:** https://pressbooks.bccampus.ca/nutr1100/chapter/defining-protein/

Even though greater than one hundred amino acids are found naturally, notably in plants, just twenty kinds of amino acids have been typically found in the majority of proteins. Peptide bonds are formed between the amino groups of 1 amino acid and the carboxyl groups of its neighboring amino acid in protein molecules, allowing the amino acids to interact with one another (Hubbard et al., 1999; Marchler-Bauer et al., 2017).

### 4.3.2. Condensation of Amino Acids

The free (—$NH_2$) (known as the N terminal of the peptide) has been on the left side of the structure, while the free (—COOH) (also known as the C terminal) has been on the right side. Proteins are polypeptides with a larger molecular weight which are huge molecules made up of numerous peptide-bonded amino acids. The majority of them have greater than one hundred amino acids bound together in a lengthy peptide chain. Every amino acid has an average molecular mass of one hundred to one hundred and twenty-five daltons (depending upon the weight of an H atom as 1), hence protein molecular weights are often in the ten thousand to one hundred thousand dalton range (1 dalton is the weight of 1 H atom). Proteins differ in their species and organ specialization due to variances in the amount and pattern of amino acids. There are 20 distinct amino acids in a 100-amino-acid chain that may be placed in over 10100 various ways (Brun et al., 2003; Fox et al., 2014).

### 4.3.3. Amino Acid Structures that are Commonly Used

The architectures of the side (R) chains of amino acids in proteins differ from one another. Glycine, which possesses R as a hydrogen atom, is the most basic amino acid. R denotes either branched or straight carbon chains in various amino acids. Alanine, which possesses the methyl group (—$CH_3$), is one of these amino acids. The alkyl side-chain sequence is completed by leucine, valine, and isoleucine, which have longer R groups. Such amino acids' alkyl side chains (R groups) are nonpolar, meaning they have little attraction to water but do have some affinity for one another. Plants can produce all alkylamino acids, however mammals can only produce alanine and glycine, hence leucine, valine, and isoleucine should all be consumed (Tatusov et al., 2001).

Serine and cysteine, 2 amino acids with 3 carbon atoms apiece, are generated from alanine. Rather than the methyl group found in alanine, serine has an alcohol group (—$CH_2OH$), while cysteine has a mercapto group (—$CH_2SH$). Serine, however not cysteine or cysteine may be synthesized by animals. Cysteine is mostly found in proteins in its oxidized state, called cystine (oxidation in this context refers to the loss of hydrogen atoms). Cystine is made up of 2 cysteine molecules joined via a disulfide bond (—S—S—) formed whenever an H atom has been extracted from every cysteine's mercapto group. Disulfide bonds have been significant in protein structure as they enable the coupling of 2 separate portions of a protein molecule to the otherwise straight chains, allowing loops to form. Small quantities of cysteine having free sulfhydryl (—SH) groups can be found in certain proteins (Zimek et al., 2008; Laibach et al., 2015).

Proteins contain 4 amino acids: asparagine, aspartic acid, methionine, and threonine each having 4 carbon atoms. Animals may produce asparagine and aspartic acid, which are abundant in nature. Because threonine and methionine may not be produced, they are necessary amino acids that should be obtained from food. Methionine is found in tiny levels in most proteins.

Proteins also comprise a 5-carbon amino acid (glutamic acid) and a secondary amine (proline), which is a ring-shaped structure wherein the amino group is connected to the alkyl side chain. Aspartic acids and Glutamic are dicarboxylic acids, which means they contain 2 carboxyl groups (—COOH).

Glutamine and asparagine are both amides of their particular dicarboxylic acid forms, meaning they contain an amide group (—$CONH_2$) instead of the side-chain carboxyl. Glutamine and glutamic acid are prevalent in major proteins; for instance, they can make up more than 1/3 of the amino acids in plant proteins. Animals may generate both glutamic acid and glutamine (Li, & Roberts, 2001; Barreto et al., 2016).

**Table 4.1.** The amino acid composition in different proteins

| Amino acid | Protein | | | | | |
|---|---|---|---|---|---|---|
| | alpha-casein | gliadin | edestin | collagen (ox hide) | keratin (wool) | myosin |
| **Histidine** | 18.7 | 11.7 | 18.6 | 4.5 | 19.7 | 15 |
| **Lysine** | 60.9 | 4.45 | 19.9 | 27.4 | 6.2 | 85 |
| **Aspartic acid**** | 63.1 | 10.1 | 99.4 | 51.9 | 51.5 | 85 |
| **Arginine** | 24.7 | 15.7 | 99.2 | 47.1 | 56.9 | 41 |
| **Serine** | 63.1 | 46.7 | 55.7 | 41.0 | 79.5 | 41 |
| **Threonine** | 41.2 | 17.6 | 31.2 | 19.3 | 55.9 | 41 |
| **Proline** | 71.3 | 117.8 | 32.9 | 125.2 | 58.3 | 22 |
| **glutamic acid**** | 153.1 | 311.0 | 144.9 | 76.2 | 99.0 | 155 |

| **Alanine** | 41.5 | 23.9 | 57.7 | 115.7 | 43.8 | 78 |
|---|---|---|---|---|---|---|
| **Glycine** | 37.3 | — | 68.0 | 354.6 | 78.0 | 39 |
| **Valine** | 53.8 | 22.7 | 54.6 | 21.4 | 46.6 | 42 |
| **half-cystine** | 3.6 | 21.3 | 10.9 | 0.0 | 105.0 | 86 |
| **Isoleucine** | 48.8 | 90.8*** | 41.9 | 14.5 | 29.0 | 42 |
| **Methionine** | 16.8 | 11.3 | 16.4 | 6.5 | 4.0 | 22 |
| **Tyrosine** | 44.7 | 17.7 | 26.9 | 5.5 | 28.7 | 18 |
| **Leucine** | 60.3 | | 60.0 | 28.2 | 59.9 | 79 |
| **Tryptophan** | 7.8 | 3.2 | 6.6 | 0.0 | 9.6 | — |
| **Phenylalanine** | 27.9 | 39.0 | 38.4 | 13.9 | 22.4 | 27 |
| **Hydroxylysine** | — | — | — | 8.0 | 1.2 | — |
| **Hydroxyproline** | 0.0 | 0.0 | 0.0 | 97.5 | 12.2 | — |
| **Total** | 839 | 765 | 883 | 1,058 | 863 | 832 |
| **Average residual weight** | 119 | 131 | 113 | 95 | 117 | 120 |

*Amount of amino acid molecules per one hundred thousand grams of protein.

**Asparagine and glutamine are included in the glutamic acid and aspartic acid levels, correspondingly.

***Isoleucine in combination with leucine.

Collagen, a protein found in animal connective tissue, contains substantial amounts of the amino acids hydroxyproline and proline. Since the amino group has been bound in a ring structure having the side chain, hydroxyproline and proline don't have free amino (—$NH_2$) groups and therefore can't exist in a zwitterion state. Whereas such amino acids' nitrogen-consisting groups (>NH) may establish peptide bonds having the carboxyl groups of other amino acids, the resulting link causes a link in the peptide chain; that is, the ring-like structure disrupts the usual bond angle of typical peptide bonds (Querol et al., 1996).

Proteins are virtually neutral molecules, which means they don't contain any basic or acidic qualities. Glutamic and Aspartic acid's acidic carboxyl (—COO−) groups are roughly equal in amount to the amino acids having primary side chains. Proteins have 3 such primary amino acids, every with 6 carbon atoms. Plants manufacture lysine, which has the simplest structure, while animals do not. Plants, too, contain a lower lysine composition. Arginine is present throughout the entire proteins, but it is especially abundant in the strongly essential protamines (basic proteins with a small number of amino acids) observed in fish sperm. Histidine is the 3rd basic amino acid. Animals may manufacture both histidine and arginine. Histidine is a less powerful amino acid than arginine or lysine. By attaching $H^+$ ions to the N atoms of the imidazole ring, a 5-membered ring structure including 2 N atoms in the side chain of histidine, the imidazole ring serves as a buffer (for example, a preservative of $H^+$ ion concentration) (Dan & Safran, 1998).

The additional tyrosine, phenylalanine, tryptophan, and amino acids, all have an aromatic structure, meaning they contain a benzene ring. All 3 amino acids have been required for life, and while mammals are unable to generate the ring of benzene, they may transform phenylalanine to tyrosine.

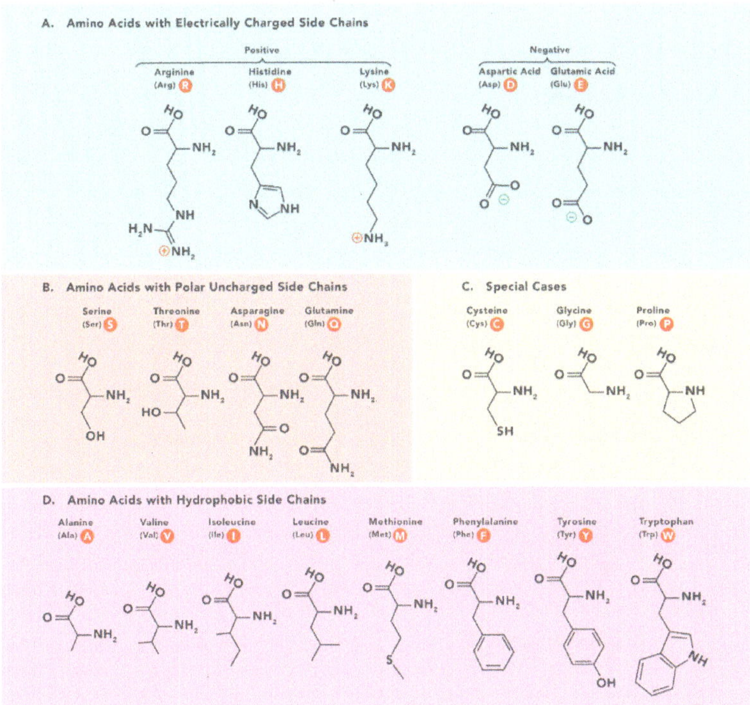


**Figure 4.5.** The twenty amino acids which make up proteins have different chemical configurations.

**Source:** https://www.technologynetworks.com/applied-sciences/articles/essential-amino-acids-chart-abbreviations-and-structure-324357

Certain amino acids catch UV light at wavelengths around 270 and 290 nanometers as they include benzene rings. Phenylalanine catches so little UV light; tryptophan and tyrosine, on either side, consume a lot of it and are accountable for the 280 to 290 nm absorption band that most proteins have.

Such absorption is frequently utilized to estimate the amount of protein in protein specimens (Hermansson, 1979; Chandrapala et al., 2011).

Usually, proteins only include the amino acids listed above; although, additional amino acids can be found in minor amounts in proteins. Collagen in connective tissue, for instance, includes modest levels of hydroxylysine in addition to hydroxyproline. Other proteins are composed of monomethyl lysine, dimethyl lysine, or trimethyl lysine, which are lysine derivatives with 1, 2, or 3 methyl groups (—$CH_3$), respectively. Furthermore, the number of such uncommon amino acids in proteins is usually more than 1% or 2% of overall amino acids (Criddle et al., 1962; Southern et al., 1991).

## 4.4. PROTEIN LEVELS OF STRUCTURAL ORGANIZATION

The four stages of organization of protein structure are primary structure, secondary structure, tertiary structure, and quaternary structure.

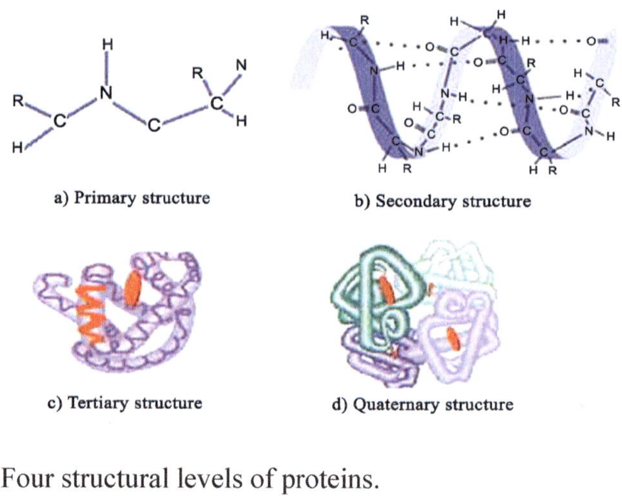


**Figure 4.6.** Four structural levels of proteins.

**Source:** https://www.researchgate.net/publication/347970038_Classifying_Protein_Sequences_Using_Convolutional_Neural_Network

### 4.4.1. Primary Structure

Protein analysis and manufacturing procedures only show the primary structure of proteins, which is the amino acid pattern of peptide chains, but not the secondary structure. They don't offer information on the peptide chain's structure (or spatial arrangement), whether it's a long straight thread or a globule that's been randomly coiled and folded which is crucial information. In interactions between amino acids, the shape, or conformation, of a protein is accessed by the repulsion or attraction of polar and nonpolar groups on the side chains (R groups) of the amino acids. Unlike the former, which carries +ve or -ve charges in its side chains, the latter is resistant to water while simultaneously attracting one another. Peptide chains comprising one hundred to two hundred amino acids can have loops or helixes, while others can be straight or create irregular coils. Certain peptide chains comprise one hundred to two hundred amino acids (Clendenning et al., 1996; Han et al., 2014).

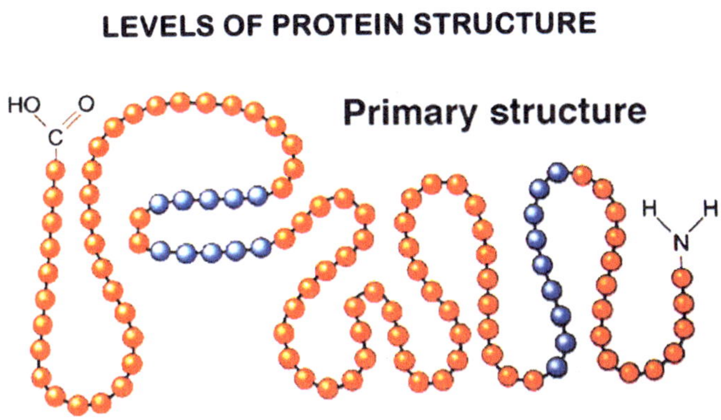


**Figure 4.7.** Protein's Primary Structure.

**Source:** https://lubrizolcdmo.com/technical-briefs/protein-structure/

Quaternary, tertiary, and secondary structures are terms used to describe the peptide chain architecture of proteins. A nomenclature committee of the IUB has expressly specified such names: Regardless of how the peptide chain is structured in space, the amino acid pattern of a protein dictates its essential structure. Regardless of the arrangement of side chains or other main chain sections, the secondary structure has been dictated only by the spatial configuration of the principal peptide chain. The tertiary structure has been dictated by both the side chains and extra surrounding sections of the major chain, with little regard for neighboring peptide chains. Ultimately, the phrase quaternary structure refers to a big protein's organization of similar or distinct subunits; each one is a single peptide chain (Salomao et al., 2008; Nekrasov et al., 2008).

### 4.4.2. Secondary Structure

Due to the size of the bond angles among consecutive atoms in a peptide chain, N and C atoms can't lay in a straight line; the bond angle is roughly 110°. Moreover, because each of the N and C atoms may spin to some amount, the chain possesses restricted elasticity. Given that all amino acids, except for glycine, are asymmetric L-amino acids, the peptide chain seems to take on an asymmetric helical shape; many fibrous proteins are composed of elongated helices arranged around a straight screw axis. These characteristics of the structure are due to traits that are shared by all peptide chains. The secondary structure of the protein is the result of such interactions (Anderson et al., 1994; Guo et al., 2021).

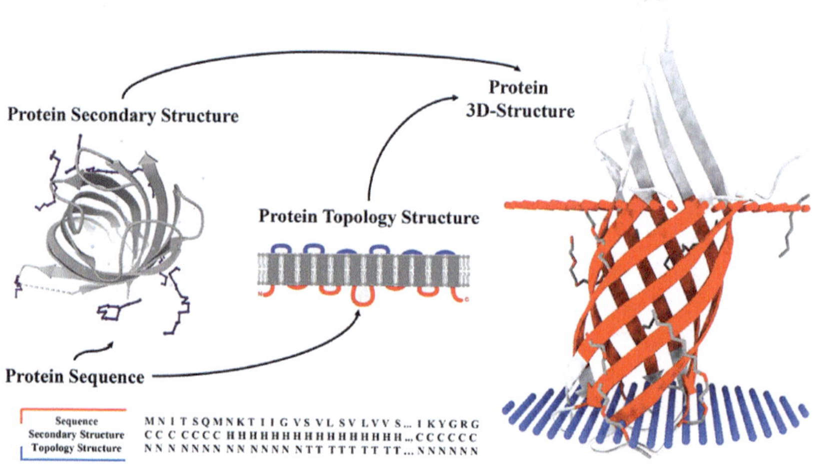


**Figure 4.8.** Protein Secondary Structure.

**Source:** Liu Z, Gong Y, Bao Y, Guo Y, Wang H and Lin GN (2021) TMPSS: A Deep Learning-Based Predictor for Secondary Structure and Topology Structure Prediction of Alpha-Helical Transmembrane Proteins. Front. Bioeng. Biotechnol. *8:629937*. doi: 10.3389/fbioe.2020.629937

### 4.4.3. Tertiary Structure

The association among the side chains (R) of the amino acids that make up the protein results in the tertiary structure. Others have been polar, while others are nonpolar. Certain of them include negatively and positively charged groups. In glycine, the amount of C atoms in the side chain is 0, while in tryptophan, it is 9. Sidechains having opposite charges pull one another, whereas side chains have the same charge oppose one another. Salt bridges are the bonds that have been produced via forces among the -ve charged side chains of glutamic acid or aspartic and the positively charged side chains of arginine or lysine on either side and the +ve charged side chains of arginine or lysine on the other side. The creation of many hydrogen bonds also contributes to the mutual attraction of nearby peptide strands (Ryazanov et al., 1987; Hanssen et al., 2001).

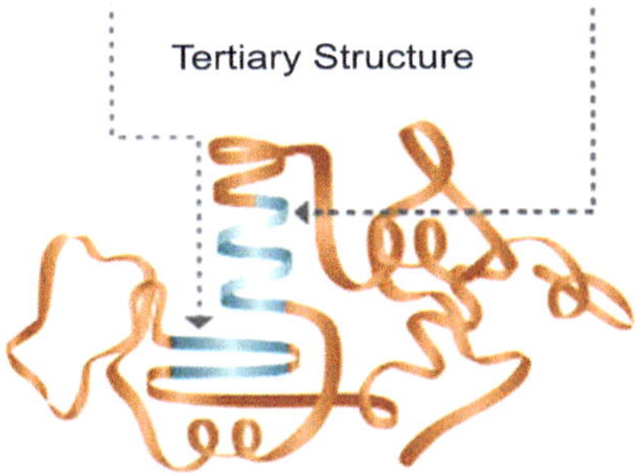


**Figure 4.9.** Protein Tertiary Structure.

**Source:** https://www.thoughtco.com/protein-structure-373563

The interaction between the N-bound H atom (amide hydrogen) and the O atom's unshared pair of electrons in the double-bonded C-O group forms H bonds (the carbonyl group). Consequently, the amide hydrogen is slightly displaced toward the carbonyl group's oxygen atom. Even though the H bond is significantly weaker as compared to a covalent bond (for example, a link between 2 C atoms wherein they possess the electrons having the same bonding pair), the huge amount of carbonyl and amide groups in peptide chains leads to the creation of multiple H bonds. The association between the nonpolar side chains of leucine, isoleucine, valine, and phenylalanine has been known as hydrophobic interaction, and it shows in the ejection of water molecules (Canu et al., 1997; Wen et al., 2005).

The disulfide bonds of cystine control the structure of the peptide chain to a large extent in cysteine-rich proteins. The cystine halves can be found at various sections of the peptide chain, forming a loop that is closed by the disulfide link.

The tertiary structure of the protein changes dramatically when the disulfide bond is decreased (for example, hydrogen is added) to 2 sulfhydryls (—SH) groups closed loops have been disrupted, and nearby disulfide-bonded peptide chains split (Björk et al., 1972; Prakash & Narasinga, 1986).

### 4.4.4. Quaternary Structure

The haemoglobin structure is a good example of the quaternary structure's characteristics. Every molecule of human haemoglobin is a tetramer, has 4 peptide chains, two α-chains, and two β-chains in every molecule. Hydrophobic associations and H bonds are responsible for the binding of the 4 subunits. Even though there are no covalent connections among the peptide chains of the 4 subunits, the haemoglobin tetramer is known as a molecule because the 4 subunits have grown to be so closely connected. Covalent bonds bind the subunits of other proteins together, preventing them from disassembling (disulfide bridges) (MacRitchie, 1992; Jia et al., 2016).

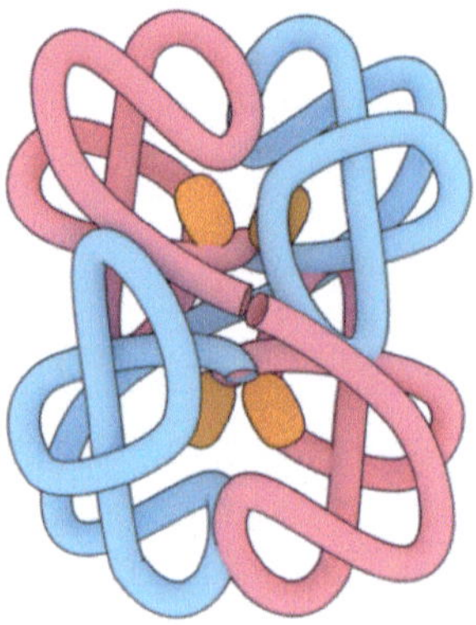

Figure 4.10. Protein Quaternary Structure

**Source:** https://commons.wikimedia.org/wiki/File:202101_Protein_quaternary_structure.svg

## 4.5. PROTEIN PHYSICOCHEMICAL PROPERTIES

### 4.5.1. Proteins' Molecular Weight

Classical chemical methods (for example, freezing-point depression) may not measure the molecular weight of proteins as they need solutions with a high protein concentration than may be generated. The lowest molecular mass of a protein or a subunit may be estimated if it includes just 1 molecule of 1 of the amino acids or 1 atom of Fe, Cu, or the other component; for instance, the protein myoglobin comprises 0.34 grams of Fe in one hundred grams of protein. Because iron has an atomic weight of 56, myoglobin's minimal molecular weight is (56 × 100)/0.34 = around about 16,500. Direct measurements of myoglobin's molecular mass provide identical results. Meanwhile, haemoglobin, which also includes 0.34 percent Fe, has a molecular weight of 66,000, or 4 × 16,500; consequently, haemoglobin comprises 4 Fe atoms (Liu et al., 2012; Theillet et al., 2014).

Ultra-centrifugation spinning in a centrifuge at rates up to sixty thousand revolutions per minute is the most common method for determining the molecular weight of proteins. At these speeds, centrifugal forces exceeding two-hundred thousand times the gravitational pull on Earth's surface are attained. In 1920, the 1st ultra-centrifuges had been constructed to find out the molecular weight of proteins. A huge amount of proteins' molecular weights are established. Most are made up of multiple subunits with a molecular weight of lower than 100,000 and a frequency of 20,000 to 30,000. Hemocyanins, the copper-consisting respiratory proteins of invertebrates, comprise proteins with extremely large molecular weights, maybe as higher as so many millions. Even though the molecular weight of proteins has no specific lower limit, small amino acid patterns have been commonly referred to as peptides (Li et al., 2007; Lu et al., 2020).

**Table 4.2.** Various Proteins' Molecular Weights

| Protein | Molecular Weight |
|---|---|
| B-Galactosidase | 116 kDa |
| Phosphorylase B | 97.4 kDa |
| Myosin | 212kDa |
| Bovine Serum Albumin | 66.2 kDa |
| Carbonic Anhydrase | 31 kDa |
| Soybean Trypsin Inhibitor | 21.4/19.7 kDa |
| Ovalbumin | 45 kDa |
| Tysozyme | 14.4 kDa |

## 4.5.2. The Shape of Protein Molecules

To do X-ray Diffraction, the X-rays have been permitted to impact a crystal of protein at an angle of 45 degrees. After passing through the crystal, the X-rays strike a photographic plate, causing a sequence of dots to appear on the image. The use of such an approach indicates that peptide chains may take on extremely intricate and seemingly irregular forms. There are 2 ends in shape: the tightly folded globular protein structure and the one-dimensional, extended structure of the fragile fibrous proteins. Both of these structures had been known several years before the development of the technology of X-ray diffraction had been discovered. The fibrous protein solutions are exceedingly thick (for example, adhesive), while the solutions of globular proteins are vicious but not sticky (such that they flow conveniently).

As an instance of how a five percent solution of the globular protein ovalbumin can flow conveniently through a glass tube, a similar five percent solution of gelatin, a fibrous protein, cannot flow through the tube even though gelatin is only liquid at an extremely higher temp and harden at ambient temperature. Although solutions consisting of only one percent or two per cent gelatin are extremely thick and flow through a tight tube either quite gradually or merely under pressure, depending on the concentration of gelatin present (Mwasaru et al., 1999; Dewald et al., 2015).

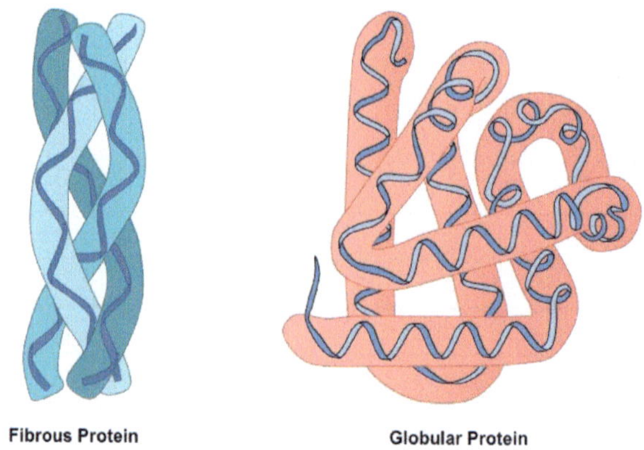


**Figure 4.11.** Globular and Fibrous Protein Structures.

**Source:** https://ib.bioninja.com.au/standard-level/topic-2-molecular-biology/24-proteins/fibrous-vs-globular-protein.html

It is possible to envisage that the extended peptide chains of the fibrous proteins become tangled not just mechanically, but also via the shared attraction of its side chains, and that this causes them to incorporate substantial quantities of $H_2O$.

Because the majority of the hydrophilic groups found in globular proteins have been found on the molecule's surface, globular proteins contain very little water by comparison. The birefringence of a solution of a fibrous protein is achieved when the solution flows through a narrow tube, with the extended molecules becoming oriented parallel to the flow direction. The solution behaves like a crystal, splitting a light ray into 2 elements that travel at varying speeds and are polarized at 90° to one another. Globular proteins, on the other hand, don't exhibit these phenomena, which is known as flow birefringence. Myosin, the tensile protein in muscles, has exceptionally higher flow birefringence; other proteins with extremely higher flow birefringence comprise fibrinogen, the clotting ingredient present in the plasma of blood, and tobacco mosaic virus solutions, among others. It has been shown that the gamma-globulins in blood plasma have little flow birefringence and that there is none to be found in solutions of ovalbumin and serum albumin (Chanana et al., 2013; Kiosseoglou et al., 2021).

## 4.6. ELECTROCHEMISTRY OF PROTEINS

There has been only 1 alpha-amino group (at the N terminal end) and 1 alpha-carboxyl group (at the C terminal end) in a particular molecule of protein since the alpha-amino and alpha-carboxyl groups of amino acids have been transformed into peptide bonds in the molecule of protein. Such 2 groups have relatively little effect on a protein's electrochemical nature. The positively charged ammonium groups of arginine and lysine, and -ve charged carboxyl groups of glutamic and aspartic acid, are important. The amount of positive and negative charged groups in most proteins ranges from ten to twenty per one hundred amino acids (Åkerström & Björck, 1989).

### 4.6.1. Electrophoresis

Because of the positive and negative charged proteins side chains, they act similarly to amino acids when exposed to electric fields; such that, they move throughout electrophoresis at the cathode (-ve terminus) having lower pH values and at the anode (+ve terminus) having higher pH values. In the case of several proteins, the isoelectric point, which is the pH value wherein the molecule of protein doesn't migrate, is in the range of pH between five to seven.

In contrast, proteins with the points of isoelectric in the pH range from eight to ten include cytochrome C, lysozyme, histone, and other proteins rich in arginine and lysine, among others. Pepsin, which includes just a small number of primary amino acids, has an isoelectric point that is near one (Garrett & Hunt, 1974; Sęczyk et al., 2019).

**Table 4.3.** The number of amino acids in a single molecule of protein

| Amino acid | Protein* | | | | | | |
|---|---|---|---|---|---|---|---|
| | **Cyto** | **Hb alpha** | **Hb beta** | **RNase** | **Lys** | **Chgen** | **Fdox** |
| **histidine** | 3 | 10 | 9 | 4 | 1 | 2 | 1 |
| **lysine** | 18 | 11 | 11 | 10 | 6 | 14 | 4 |
| **aspartic acid**** | 8 | 12 | 13 | 15 | 21 | 23 | 13 |
| **arginine** | 2 | 3 | 3 | 4 | 11 | 4 | 1 |
| **serine** | 2 | 11 | 5 | 15 | 10 | 28 | 7 |
| **threonine** | 7 | 9 | 7 | 10 | 7 | 23 | 8 |
| **proline** | 4 | 7 | 7 | 4 | 2 | 9 | 4 |
| **glutamic acid**** | 10 | 5 | 11 | 12 | 5 | 15 | 13 |
| **alanine** | 6 | 21 | 15 | 12 | 12 | 22 | 9 |
| **glycine** | 13 | 7 | 13 | 3 | 12 | 23 | 6 |
| **valine** | 3 | 13 | 18 | 9 | 6 | 23 | 7 |
| **half-cystine** | 2 | 1 | 2 | 8 | 8 | 10 | 5 |
| **isoleucine** | 8 | 0 | 0 | 3 | 6 | 10 | 4 |
| **methionine** | 3 | 2 | 1 | 4 | 2 | 2 | 0 |
| **leucine** | 6 | 18 | 18 | 2 | 8 | 19 | 8 |
| **tryptophan** | 1 | 1 | 2 | 0 | 6 | 8 | 1 |
| **tyrosine** | 5 | 3 | 3 | 6 | 3 | 4 | 4 |
| **phenylalanine** | 3 | 7 | 8 | 3 | 3 | 6 | 2 |
| **total** | 104 | 141 | 146 | 124 | 129 | 245 | 97 |

*Hb alpha = human hemoglobin A, alpha-chain; Cyto = human cytochrome c; Hb beta = human hemoglobin A, beta-chain; Lys = chicken lysozyme;

RNase = bovine ribonuclease; Fdox = spinach ferredoxin; Chgen = bovine chymotrypsinogen.

**Glutamine and asparagine are included in the numbers for glutamic acid and aspartic acid, correspondingly.

Free-boundary electrophoresis, the authentic technique for accessing electrophoretic migration, is largely replaced by zone electrophoresis, wherein the protein has been positioned in a gel of polyacrylamide, agar, or starch, or a permeable material like cellulose or paper acetate, before being separated by a gel of sodium chloride or potassium chloride. Visual observation of the movement of hemoglobin and other colored proteins is possible. To make colorless proteins visible following electrophoresis, a suitable dye is used to stain them once the process is complete (Da C Azeredo et al., 2003; Zhou et al., 2017).

## 4.6.2. The Chemical Reactivity of Proteins

Chemical approaches may indicate if particular groups are existed on the surface of the protein molecule and therefore capable to respond, or if they have been buried within the tightly folded peptide chains and therefore have been not able to respond, providing information on the interior structure of proteins. The chemical reagents utilized in these studies should be non-toxic and do not affect the protein's structure (Sarda et al., 2005; Jiang et al., 2017).

The reactivity of tyrosine is particularly fascinating. For instance, only 3 of the 6 tyrosines identified in the naturally present enzyme ribonuclease may be iodinated, according to research (such as reacted to accept an iodine atom). The peptides containing iodinated tyrosines have been identified using an enzyme-catalyzed breakdown of iodinated ribonuclease. The 3 tyrosines which may be iodinated have been found on the ribonuclease surface; the others have been thought to be buried deep within the molecule. Other approaches, such as treatment using diazonium compounds or tetranitromethane, may be utilized to identify tyrosine. Whenever the protein is broken down by enzymes, the colored chemicals that arise may easily be recognized.

Cysteine may be identified by combining it with iodoacetic acid or iodoacetamide; the reaction produces carboxymethyl L-Cysteine or carbamidomethyl L-Cysteine, which may be identified via measuring the amino acid content of peptides including them. Some histidines' imidazole groups may be found by combining with identical reagents under different circumstances. However, few additional amino acids may be labeled without causing modifications to the protein's secondary or tertiary structure (Guo et al., 1991; Ferapontova et al., 2005).

## 4.7. STRUCTURAL PROTEINS

### 4.7.1. Collagen

The structural protein collagen has been found in tendons, bones, skin, and ligaments. For a long time, collagen was thought not to be soluble in water. Furthermore, a portion of the collagen in calfskin may be removed using a citrate buffer at pH 3.7. Procollagen, a collagen precursor, is transformed into collagen in the body. The molecular weight of procollagen is 120,000. Collagen is formed by cleaving 1 or certain peptide bonds in procollagen, yielding 3 subunits, including a molecular mass of 95,000; hence, the collagen molecular mass is 285,000 (3 × 95,000). The 3 subunits have been spiral-wounded around an extended, straight axis. Every subunit has a length of 2,900 angstroms and a diameter of around fifteen angstroms. Because the 3 chains have been scattered, the trimer has no clear endpoints (Stankovich & Bard, 1977; Arrigan et al., 2018).

Collagen is unique among proteins in that it contains a large amount of hydroxyproline and proline. Apart from elastin, hydroxyproline is not found in substantial concentrations in some other proteins. The majority of proline in collagen is found in the glycine–proline–X pattern, where X is typically hydroxyproline or alanine. Because collagen lacks tryptophan and cysteine, it may not be used to replace other proteins in the diet. Proline produces bends in the peptide chain, reducing the amino acid unit's length between 3.7 angstroms in the elongated chain of the structure and 2.86 angstroms in the collagen chain. The glycines are inside the tangled triple helix, near the axis, while the prolines are on the outside (Stankovich & Bard, 1978; Suprun, 2021).

The bacterial enzyme collagenase hydrolyzes native collagen, which is resistant to trypsin. The triple helix is disrupted and the subunits are partly hydrolyzed when collagen is boiled with water; the result is gelatin. Gelatin's unwound peptide chains capture a lot of water, resulting in a hydrated molecule.

Whenever the collagen has been processed with chromium salts or tannic acid, cross-links develop amongst the fibers of collagen, rendering it intractable; such tanning procedure is used to turn to hide into leather. The tanned substance may not be turned into gelatin because it is insoluble in hot water. The cross-links created by the tanning chemicals break down when exposed to water at sixty-two degrees Celcius to sixty-three degrees Celcius, and the leather compresses permanently to roughly one-third of its initial volume.

Collagen appears to age in living beings, which might be due to the creation of cross-links among collagen fibers. They have been made by converting certain lysine side chains to aldehydes (compounds having the common formula RCHO) and then combining the aldehydes having intact lysine side chains' ε-amino groups. Identical cross-links can be seen in the protein elastin which is found in connective tissue elastic fibers and can develop through the interaction of collagen fibers with other proteins. Isodesmosins and desmosine are results of cross-linked lysine fragments generated while cross-linked elastin or collagen is degraded (Stankovich & Bard, 1977; Zanello, 2019).

## 4.7.2. Keratin

Keratin is a structural protein found in epithelial cells in the outer layers of the skin and is found in hair, hooves, nails, and feathers. Keratin is not entirely soluble in either warm or cold water, and this has been unaffected by proteolytic enzymes, hence it may not be used as a protein replacement in the diet. Keratin's exceptional stability is due to cystine's many disulfide bonds. Its amino acid content varies from collagen. Cystine can account for up to a quarter of all amino acids in the body. The peptide chains in keratin have been assembled in nearly equal quantities of parallel and unparallel pleated sheets, with H bonds connecting the amino and carbonyl groups (BURROWS et al., 1991; Bowden, 1997).

The disulfide bonds are broken down into sulfhydryl groups, leading to the disintegration of the peptide chains, which contains a molecular mass ranging from twenty-five grams to twenty-eight grams each. Using thioglycolic acid or a similar mild reducing agent, hair keratin is partially disulfide bond reduced, followed through oxidation of the hair's reoriented (—SH) to (—S—S—), which occurs when the hair is exposed to oxygen in the atmosphere.

The quantity of water contained inside fibers of keratin determines the length of the fibers. They can bind up to sixteen percent of water, and such hydration has been followed by a ten percent to twelve percent rise in the length of the fibers themselves. The most extensive study has been conducted on keratin from hair, particularly wool keratin. It is composed of a mixture of cysteine-rich and cystine-deficient peptides. When wool is warmed in water to around ninety degrees Celcius, it shrinks irreversibly. According to current thinking, the reason is the disintegration of H bonds and other non-covalent bonds; (—S—S—) bonds appear to be unaffected. (Gruenbaum et al., 2000)

## 4.8. PROTEIN HORMONES

Some of the hormones generated by the endocrine glands are peptides or proteins, while others may be steroids. Hormones' origins, physiological roles, and mechanisms of action are covered under the article hormone. There is no enzymatic activity in any of the hormones. Everyone contains a target organ to which it induces a biological reaction, such as pancreatic or gastric juice secretion, milk production, or steroid hormone production. Hormones have an unknown mechanism for exerting their effects. The hormone sends a signal to the cells through cyclic adenosine monophosphate, which boosts the activity of the cells (Wistow & Piatigorsky, 1987).

### 4.8.1. Hormones of the Thyroid Gland

Thyroglobulin, whose active groups comprise 2 molecules of the iodine-containing chemical thyroxine, contains a molecular mass of 670,000 and is composed of 2 two molecules of thyroxine. Thyroglobulin also includes thyroxine containing 2 and 3 iodine atoms rather than 4, and tyrosine including 1 and 2 iodine atoms rather than 2. The hormone injection induces an increase in metabolism, whereas the absence of the hormone causes a slowdown. The thyroid gland also produces another hormone, calcitonin, which has the effect of lowering the calcium level in the blood. Certain amino acids in the amino acid pattern of calcitonin from beef, and salmon vary from the amino acid pattern of human calcitonin. The whole of them, on either hand, contain the half-cystines (C) and the prolinamide (P) in identical positions to each other (Hatzfeld, 1998).

To keep the calcium level in the blood stable, parathyroid hormone (parathormone), which is generated by tiny glands that have been embedded inside or concealed in the thyroid gland, must be present in the body. Hypocalcemia is caused by a decrease in the amount of calcium produced (a decrement in calcium levels in the bloodstream underneath the usual range).

This hormone has a molecular mass of 8,500; it has no cysteine or cystine and is high in glutamic acid, aspartic acid, or its amides, but it does not have any cysteine (Presland & Dale, 2000).

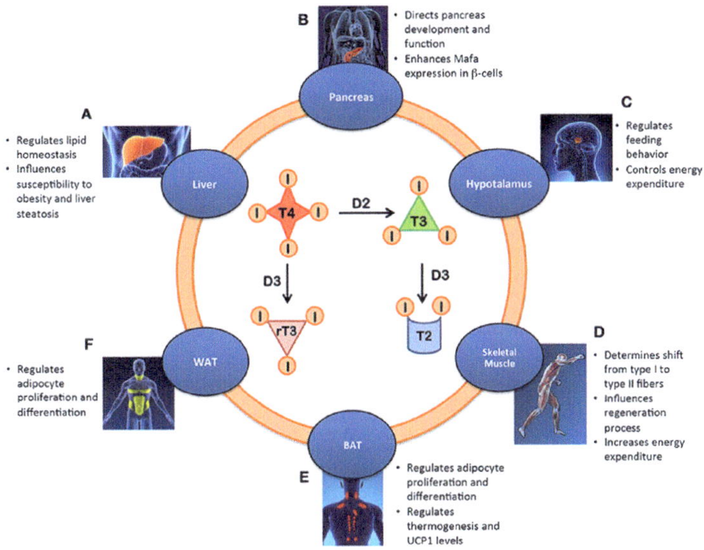


**Figure 4.12.** Thyroid Hormones.

**Source:** Cicatiello AG, Di Girolamo D and Dentice M (2018) Metabolic Effects of the Intracellular Regulation of Thyroid Hormone: Old Players, New Concepts. Front. Endocrinol. 9:474. doi: 10.3389/fendo.2018.00474

## 4.8.2. Pancreatic Hormones

Even though the insulin‘s amino acid structure is recognized since 1949, efforts to manufacture it have shown unsatisfactory results due to the inability of the 2 peptide chains to join and create the proper disulfide bridge. The finding of proinsulin, from which insulin is generated; the pancreas explains the simplicity with which insulin is biosynthesized. Proinsulin's single peptide chain loses a 33-amino-acid peptide known as the C peptide, or linking peptide, throughout its insulin conversion. Proinsulin's disulfide bridges join the A and B chains (Laemmli, 1970; Keller, 1993).

Insulin is mostly found in aqueous solutions as a combination of 6 subunits, each with an A and B chain. Insulins from many species are extracted and examined; their amino acid pattern change somewhat, but they all feature identical disulfide bridges connecting the 2 chains. Even though the infusion of insulin decreases the sugar in the blood, the glucagon administration, another pancreatic hormone, elevates it. A 29-amino-acid straight peptide chain makes up glucagon. It has been manufactured, and it possesses all of the biological activity of natural glucagon. Glucagon has neither cystine nor isoleucine in its structure (Spear & Roizman, 1972; De Vries et al., 1992 ).

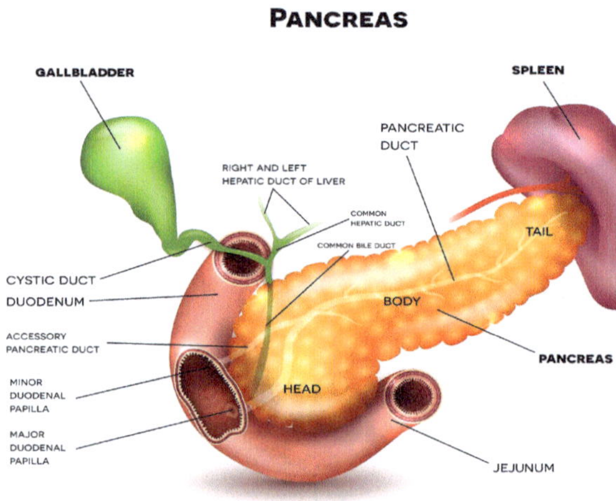


**Figure 4.13.** General structure of the pancreas.

**Source:** Karpińska M and Czauderna M (2022) Pancreas—Its Functions, Disorders, and Physiological Impact on the Mammals' Organism. Front. Physiol. *13:807632*. doi: 10.3389/fphys.2022.807632

It is divided into three parts: a posterior lobe, an anterior lobe, and an intermediary section; they vary in cell structure as well as the structure and function of the hormones that are produced by each part of the pituitary gland. The posterior lobe is responsible for the production of 2 hormones that are quite similar: oxytocin and vasopressin. The former generates contractions in the uterus of a pregnant woman, while the latter increases blood pressure. These 2 amino acids are octapeptides, which are made up of a ring of 5 amino acids (the 2 cystine halves counted like one amino acid each) and a side chain of 3 amino acids each. Disulfide bonds connect the 2 cystine halves, and glycinamide is the amino acid at the C terminal of the cysteine molecule. An established and confirmed structure is constructed and established. Human vasopressin varies from oxytocin in that the amino acid isoleucine has been replaced via phenylalanine, and the amino acid leucine has been substituted by arginine (Brown, 1952; Brisson et al., 1994). The melanocyte-stimulating hormone (MSH) is formed by the intermediary section of the pituitary gland, which induces the growth of pigmented melanophores (cells) in the frog's skin and other batrachians. Both hormones, referred to as α-MSH and β-MSH, are synthesized from hog pituitary glands, respectively. There are two types of amino acids in this category: the 1st, α-MSH, has thirteen amino acids; its N terminus serine has been acetylated (for example the acetyl group, $CH_3CO$, of acetic acid has been attached), and the second contains the amino acid valinamide at its C terminus valine residue. The 2nd compound, β-MSH, has several of the amino acids found in α-MSH in its eighteen amino acids (Debrunner & Frauenfelder, 1982).

Thyroid-stimulating hormone (thyrotropin), which has a molecular weight of 28,000, a lactogenic hormone, which contains a molecular weight of 22,500, a growth hormone, which contains a molecular weight of 21,500, a luteinizing hormone, which has a molecular weight of 30,000, and a follicle-stimulating hormone, which has a molecular weight of 29,000, are all produced by the anterior Thyroid-stimulating hormone is made up of two subunits, and, which are similar to the luteinizing hormone subunits in terms of composition. When the two subunits are separated, neither has any hormonal function; however, when they are reconnected, they resume about half of their previous activity. Prolactin is a lactogenic hormone produced from sheep pituitary glands with a total of 190 amino acids. Researchers created an identical peptide chain of one hundred and eighty-eight amino acids based on their pattern, which not only includes ten percent of the biological activity of the naturally occurring hormone but also ranks among the activities of growth hormone. The growth hormone (somatotropic hormone) has an amino acid pattern that is identified; this hormone appears to boost RNA production and, as a result, accelerates the rate of growth in the body. In the body, the luteinizing hormone is found in mucoproteins that include roughly 12% carbohydrates. It is composed of 2 subunits, each of which has a molecular weight in the range of 15,000; when divided, the subunits autonomously rejoin. Chorionic gonadotropin is found in the urine of pregnant women, and the existence of this hormone allows for the detection of pregnancy at a preliminary phase. The amino acid pattern has been determined. Its amino acid sequence is similar to that of the growth hormone in 160 of its 190 amino acids, and one hundred such amino acids are also found in a similar pattern as those found in lactogenic hormones. It is possible that the various pituitary hormones, as well as the chorionic gonadotropin, were generated in this manner (Matsuura & Yomo, 2006; Xie et al., 2007).

# REFERENCES

1. Åkerström, B., & Björck, L. (1989). Protein L: an immunoglobulin light chain-binding bacterial protein: characterization of binding and physicochemical properties. *Journal of Biological Chemistry*, *264*(33), 19740-19746.
2. Anderson, J. V., Li, Q. B., Haskell, D. W., & Guy, C. L. (1994). Structural organization of the spinach endoplasmic reticulum-luminal 70-kilodalton heat-shock cognate gene and expression of 70-kilodalton heat-shock genes during cold acclimation. *Plant Physiology*, *104*(4), 1359-1370.
3. Arrigan, D. W., Hackett, M. J., & Mancera, R. L. (2018). Electrochemistry of proteins at the interface between two immiscible electrolyte solutions. *Current Opinion in Electrochemistry*, *12*, 27-32.
4. Barreto, S. C., Ray, A., & Ag Edgar, P. (2016). Biological characteristics of CCN proteins in tumor development. *J BUON*, *21*(6), 1359-1367.
5. Björk, I., Petersson, B. Å., & Sjöquist, J. (1972). Some physicochemical properties of protein A from Staphylococcus aureus. *European Journal of Biochemistry*, *29*(3), 579-584.
6. Bowden, E. F. (1997). Wiring Mother Nature (Interfacial electrochemistry of proteins.). *The Electrochemical Society interface*, *6*(4), 40.
7. Brisson, L. F., Tenhaken, R., & Lamb, C. (1994). Function of oxidative cross-linking of cell wall structural proteins in plant disease resistance. *The Plant Cell*, *6*(12), 1703-1712.
8. Brown, C. H. (1952). Some structural proteins of *Mytilus edulis*. *Journal of Cell Science*, *3*(24), 487-502.
9. Brun, C., Chevenet, F., Martin, D., Wojcik, J., Guénoche, A., & Jacq, B. (2003). Functional classification of proteins for the prediction of cellular function from a protein-protein interaction network. *Genome biology*, *5*(1), 1-13.
10. BURROWS, A. L., GUO, L. H., HILL, H. A. O., McLENDON, G., & SHERMAN, F. (1991). Direct electrochemistry of proteins: Investigations of yeast cytochrome c mutants and their complexes with cytochrome b5. *European journal of biochemistry*, *202*(2), 543-549.
11. Canu, N., Possenti, R., Ricco, A. S., Rocchi, M., & Levi, A. (1997). Cloning, structural organization analysis, and chromosomal assignment of the human gene for the neurosecretory protein VGF. *Genomics*, *45*(2), 443-446.

12. Chanana, M., Rivera_Gil, P., Correa-Duarte, M. A., Liz-Marzán, L. M., & Parak, W. J. (2013). Physicochemical properties of protein-coated gold nanoparticles in biological fluids and cells before and after proteolytic digestion. *Angewandte Chemie International Edition*, *52*(15), 4179-4183.
13. Chandrapala, J., Zisu, B., Palmer, M., Kentish, S., & Ashokkumar, M. (2011). Effects of ultrasound on the thermal and structural characteristics of proteins in reconstituted whey protein concentrate. *Ultrasonics sonochemistry*, *18*(5), 951-957.
14. Clendenning, J. B., Humbert, R., Green, E. D., Wood, C., Traver, D., & Furlong, C. E. (1996). Structural organization of the HumanPON1Gene. *Genomics*, *35*(3), 586-589.
15. Cohn, E. J. (1941). The properties and functions of the plasma proteins, with a consideration of the methods for their separation and purification. *Chemical Reviews*, *28*(2), 395-417.
16. Criddle, R. S., Bock, R. M., Green, D. E., & Tisdale, H. (1962). Physical characteristics of proteins of the electron transfer system and interpretation of the structure of the mitochondrion. *Biochemistry*, *1*(5), 827-842.
17. Da C Azeredo, L., Azeredo, M. A. A., De Souza, S. R., & Dutra, V. M. L. (2003). Protein contents and physicochemical properties in honey samples of Apis mellifera of different floral origins. *Food chemistry*, *80*(2), 249-254.
18. Dan, N., & Safran, S. A. (1998). Effect of lipid characteristics on the structure of transmembrane proteins. *Biophysical journal*, *75*(3), 1410-1414.
19. De Vries, A. A., Chirnside, E. D., Horzinek, M. C., & Rottier, P. J. (1992). Structural proteins of equine arteritis virus. *Journal of virology*, *66*(11), 6294-6303.
20. Debrunner, P. G., & Frauenfelder, H. (1982). Dynamics of proteins. *Annual review of physical chemistry*, *33*(1), 283-299.
21. Dewald, I., Isakin, O., Schubert, J., Kraus, T., & Chanana, M. (2015). Protein identity and environmental parameters determine the final physicochemical properties of protein-coated metal nanoparticles. *The Journal of Physical Chemistry C*, *119*(45), 25482-25492.
22. Ferapontova, E. E., Shleev, S., Ruzgas, T., Stoica, L., Christenson, A., Tkac, J., ... & Gorton, L. (2005). Direct electrochemistry of proteins

and enzymes. In *Perspectives in bioanalysis* (Vol. 1, pp. 517-598). Elsevier.

23. Fox, N. K., Brenner, S. E., & Chandonia, J. M. (2014). SCOPe: Structural Classification of Proteins—extended, integrating SCOP and ASTRAL data and classification of new structures. *Nucleic acids research*, *42*(D1), D304-D309.
24. Garrett, E. R., & Hunt, C. A. (1974). Physicochemical properties, solubility, and protein binding of Δ9-tetrahydrocannabinol. *Journal of pharmaceutical sciences*, *63*(7), 1056-1064.
25. Gosline, J., Lillie, M., Carrington, E., Guerette, P., Ortlepp, C., & Savage, K. (2002). Elastic proteins: biological roles and mechanical properties. *Philosophical Transactions of the Royal Society of London. Series B: Biological Sciences*, *357*(1418), 121-132.
26. Gruenbaum, Y., Wilson, K. L., Harel, A., Goldberg, M., & Cohen, M. (2000). Nuclear lamins—structural proteins with fundamental functions. *Journal of structural biology*, *129*(2-3), 313-323.
27. Guo, L. H., Allen, H., & Hill, O. (1991). Direct electrochemistry of proteins and enzymes. In *Advances in inorganic chemistry* (Vol. 36, pp. 341-375). Academic Press.
28. Guo, M., Zhao, L., Liu, J., Wang, X., Yao, H., Chang, X., ... & Chen, C. (2021). The underlying function and structural organization of the intracellular protein corona on graphdiyne oxide nanosheet for local immunomodulation. *Nano Letters*, *21*(14), 6005-6013.
29. Habermann, B. (2004). The BAR-domain family of proteins: a case of bending and binding? The membrane bending and GTPase-binding functions of proteins from the BAR-domain family. *EMBO reports*, *5*(3), 250-255.
30. Han, M., Wang, P., Xu, X., & Zhou, G. (2014). Low-field NMR study of heat-induced gelation of pork myofibrillar proteins and its relationship with microstructural characteristics. *Food Research International*, *62*, 1175-1182.
31. Hanssen, E., Franc, S., & Garrone, R. (2001). Synthesis and structural organization of zonular fibers during development and aging. *Matrix Biology*, *20*(2), 77-85.
32. Hatzfeld, M. (1998). The armadillo family of structural proteins. *International review of cytology*, *186*, 179-224.
33. Hermansson, A. M. (1979). Methods of studying functional

characteristics of vegetable proteins. *Journal of the American Oil Chemists' Society*, *56*(3Part2), 272-279.

34. Hu, L., Huang, T., Shi, X., Lu, W. C., Cai, Y. D., & Chou, K. C. (2011). Predicting functions of proteins in mouse based on weighted protein-protein interaction network and protein hybrid properties. *PloS one*, *6*(1), e14556.
35. Hubbard, T. J., Ailey, B., Brenner, S. E., Murzin, A. G., & Chothia, C. (1999). SCOP: a Structural Classification of Proteins database. *Nucleic acids research*, *27*(1), 254-256.
36. Hubbard, T. J., Murzin, A. G., Brenner, S. E., & Chothia, C. (1997). SCOP: a structural classification of proteins database. *Nucleic acids research*, *25*(1), 236-239.
37. Isralewitz, B., Gao, M., & Schulten, K. (2001). Steered molecular dynamics and mechanical functions of proteins. *Current opinion in structural biology*, *11*(2), 224-230.
38. Jia, J., Liu, Z., Xiao, X., Liu, B., & Chou, K. C. (2016). Identification of protein-protein binding sites by incorporating the physicochemical properties and stationary wavelet transforms into pseudo amino acid composition. *Journal of Biomolecular Structure and Dynamics*, *34*(9), 1946-1961.
39. Jiang, S., Ding, J., Andrade, J., Rababah, T. M., Almajwal, A., Abulmeaty, M. M., & Feng, H. (2017). Modifying the physicochemical properties of pea protein by pH-shifting and ultrasound combined treatments. *Ultrasonics sonochemistry*, *38*, 835-842.
40. Keller, B. (1993). Structural cell wall proteins. *Plant Physiology*, *101*(4), 1127.
41. Kiosseoglou, V., Paraskevopoulou, A., & Poojary, M. M. (2021). Functional and physicochemical properties of pulse proteins. In *Pulse foods* (pp. 113-146). Academic Press.
42. LAEMMLI, U. K. (1970). Cleavage of structural proteins during the assembly of the head of bacteriophage T4. *nature*, *227*(5259), 680-685.
43. Laibach, N., Post, J., Twyman, R. M., Gronover, C. S., & Prüfer, D. (2015). The characteristics and potential applications of structural lipid droplet proteins in plants. *Journal of Biotechnology*, *201*, 15-27.
44. Li, D., & Roberts, R. (2001). Human Genome and Diseases:¶ WD-repeat proteins: structure characteristics, biological function, and their involvement in human diseases. *Cellular and Molecular Life Sciences CMLS*, *58*(14), 2085-2097.

45. Li, Y., Chen, Z., & Mo, H. (2007). Effects of pulsed electric fields on physicochemical properties of soybean protein isolates. *LWT-Food Science and Technology*, *40*(7), 1167-1175.
46. Liu, J., Ru, Q., & Ding, Y. (2012). Glycation a promising method for food protein modification: Physicochemical properties and structure, a review. *Food Research International*, *49*(1), 170-183.
47. Lo Conte, L., Ailey, B., Hubbard, T. J., Brenner, S. E., Murzin, A. G., & Chothia, C. (2000). SCOP: a structural classification of proteins database. *Nucleic acids research*, *28*(1), 257-259.
48. Lu, Z. X., He, J. F., Zhang, Y. C., & Bing, D. J. (2020). Composition, physicochemical properties of pea protein and its application in functional foods. *Critical Reviews in Food Science and Nutrition*, *60*(15), 2593-2605.
49. MacRitchie, F. (1992). Physicochemical properties of wheat proteins in relation to functionality. In *Advances in food and nutrition research* (Vol. 36, pp. 1-87). Academic Press.
50. Marchler-Bauer, A., Bo, Y., Han, L., He, J., Lanczycki, C. J., Lu, S., ... & Bryant, S. H. (2017). CDD/SPARCLE: functional classification of proteins via subfamily domain architectures. *Nucleic acids research*, *45*(D1), D200-D203.
51. Matsuura, T., & Yomo, T. (2006). In vitro evolution of proteins. *Journal of bioscience and bioengineering*, *101*(6), 449-456.
52. Murzin, A. G. (1996). Structural classification of proteins: new superfamilies. *Current opinion in structural biology*, *6*(3), 386-394.
53. Murzin, A. G., Brenner, S. E., Hubbard, T., & Chothia, C. (1995). SCOP: a structural classification of proteins database for the investigation of sequences and structures. *Journal of molecular biology*, *247*(4), 536-540.
54. Mwasaru, M. A., Muhammad, K., Bakar, J., & Man, Y. B. C. (1999). Effects of isolation technique and conditions on the extractability, physicochemical and functional properties of pigeonpea (*Cajanus cajan*) and cowpea (*Vigna unguiculata*) protein isolates. I. Physicochemical properties. *Food chemistry*, *67*(4), 435-443.
55. Nekrasov, A. N., Anashkina, A. A., & Zinchenko, A. A. (2014). A new paradigm of protein structural organization. In *Proceedings of the 2nd International Conference "Theoretical Approaches to BioInformation Systems"(TABIS 2013) Institute of Physics Belgrade, Serbia* (pp. 1-22).

56. Prakash, V., & Narasinga Rao, M. S. (1986). Physicochemical properties of oilseed protein. *Critical Reviews in Biochemistry*, *20*(3), 265-363.

57. Presland, R. B., & Dale, B. A. (2000). Epithelial structural proteins of the skin and oral cavity: function in health and disease. *Critical Reviews in Oral Biology & Medicine*, *11*(4), 383-408.

58. Querol, E., Perez-Pons, J. A., & Mozo-Villarias, A. (1996). Analysis of protein conformational characteristics related to thermostability. *Protein Engineering, Design and Selection*, *9*(3), 265-271.

59. Roitberg, A., Gerber, R. B., Elber, R., & Ratner, M. A. (1995). Anharmonic wave functions of proteins: quantum self-consistent field calculations of BPTI. *Science*, *268*(5215), 1319-1322.

60. Ryazanov, A. G., Ovchinnikov, L. P., & Spirin, A. S. (1987). Development of structural organization of protein-synthesizing machinery from prokaryotes to eukaryotes. *Biosystems*, *20*(3), 275-288.

61. Salomao, M., Zhang, X., Yang, Y., Lee, S., Hartwig, J. H., Chasis, J. A., ... & An, X. (2008). Protein 4.1 R-dependent multiprotein complex: new insights into the structural organization of the red blood cell membrane. *Proceedings of the National Academy of Sciences*, *105*(23), 8026-8031.

62. Sarda, D., Chua, G. H., Li, K. B., & Krishnan, A. (2005). pSLIP: SVM based protein subcellular localization prediction using multiple physicochemical properties. *Bmc Bioinformatics*, *6*(1), 1-12.

63. Sarkar, S. N., & Sen, G. C. (2004). Novel functions of proteins encoded by viral stress-inducible genes. *Pharmacology & therapeutics*, *103*(3), 245-259.

64. Schwartz, J. C., Cech, T. R., & Parker, R. R. (2015). Biochemical properties and biological functions of FET proteins. *Annual review of biochemistry*, *84*, 355-379.

65. Sęczyk, Ł., Świeca, M., Kapusta, I., & Gawlik-Dziki, U. (2019). Protein–phenolic interactions as a factor affecting the physicochemical properties of white bean proteins. *Molecules*, *24*(3), 408.

66. Southern, J. A., Young, D. F., Heaney, F., Baumgärtner, W. K., & Randall, R. E. (1991). Identification of an epitope on the P and V proteins of simian virus 5 that distinguishes between two isolates with different biological characteristics. *Journal of General Virology*, *72*(7), 1551-1557.

67. Spear, P. G., & Roizman, B. (1972). Proteins specified by herpes simplex virus V. Purification and structural proteins of the herpesvirion. *Journal of virology*, *9*(1), 143-159.
68. Stankovich, M. T., & Bard, A. J. (1977). The electrochemistry of proteins and related substances: Part II. Insulin. *Journal of Electroanalytical Chemistry and Interfacial Electrochemistry*, *85*(1), 173-183.
69. Stankovich, M. T., & Bard, A. J. (1977). The electrochemistry of proteins and related substances: I. Cystine and cysteine at the mercury electrode. *Journal of Electroanalytical Chemistry and Interfacial Electrochemistry*, *75*(2), 487-505.
70. Stankovich, M. T., & Bard, A. J. (1978). The electrochemistry of proteins and related substances part III. Bovine serum albumin. *Journal of Electroanalytical Chemistry and Interfacial Electrochemistry*, *86*(1), 189-199.
71. Suprun, E. V. (2021). Direct electrochemistry of proteins and nucleic acids: The focus on 3D structure. *Electrochemistry Communications*, *125*, 106983.
72. Tatusov, R. L., Natale, D. A., Garkavtsev, I. V., Tatusova, T. A., Shankavaram, U. T., Rao, B. S., ... & Koonin, E. V. (2001). The COG database: new developments in phylogenetic classification of proteins from complete genomes. *Nucleic acids research*, *29*(1), 22-28.
73. Theillet, F. X., Binolfi, A., Frembgen-Kesner, T., Hingorani, K., Sarkar, M., Kyne, C., ... & Selenko, P. (2014). Physicochemical properties of cells and their effects on intrinsically disordered proteins (IDPs). *Chemical reviews*, *114*(13), 6661-6714.
74. Vestal, D. J., & Jeyaratnam, J. A. (2011). The guanylate-binding proteins: emerging insights into the biochemical properties and functions of this family of large interferon-induced guanosine triphosphatase. *Journal of Interferon & Cytokine Research*, *31*(1), 89-97.
75. Wen, X., Lei, Y. P., Zhou, Y. L., Okamoto, C. T., Snead, M. L., & Paine, M. L. (2005). Structural organization and cellular localization of tuftelin-interacting protein 11 (TFIP11). *Cellular and Molecular Life Sciences CMLS*, *62*(9), 1038-1046.
76. Wistow, G., & Piatigorsky, J. (1987). Recruitment of enzymes as lens structural proteins. *Science*, *236*(4808), 1554-1556.
77. Xie, H., Vucetic, S., Iakoucheva, L. M., Oldfield, C. J., Dunker, A. K., Uversky, V. N., & Obradovic, Z. (2007). Functional anthology of

intrinsic disorder. 1. Biological processes and functions of proteins with long disordered regions. *Journal of proteome research*, *6*(5), 1882-1898.

78. Zanello, P. (2019). Structure and electrochemistry of proteins harboring iron-sulfur clusters of different nuclearities. Part V. Nitrogenases. *Coordination Chemistry Reviews*, *398*, 113004.
79. Zhou, X., Jiang, S., Zhao, D., Zhang, J., Gu, S., Pan, Z., & Ding, Y. (2017). Changes in physicochemical properties and protein structure of surimi enhanced with camellia tea oil. *Lwt*, *84*, 562-571.
80. Zimek, A., Buchwald, F., Frank, E., & Kramer, S. (2008). A study of hierarchical and flat classification of proteins. *IEEE/ACM Transactions on Computational Biology and Bioinformatics*, *7*(3), 563-571.
81. Zoraghi, R., Corbin, J. D., & Francis, S. H. (2004). Properties and functions of GAF domains in cyclic nucleotide phosphodiesterases and other proteins. *Molecular pharmacology*, *65*(2), 267-278.

CHAPTER 5

# ROLE OF DNA AND CHROMOSOMES IN CELL

CONTENTS

## 5.1. INTRODUCTION

Genomics refers to genetic information of DNA that provides the commands for constructing a particular protein that interacts with single or many kinds of cells in the body. Chromosomes are biological structures that contain the genetic information of a person. Genes are carried by chromosomes, which have been discovered in the nucleus of the cell. On one chromosome, there have been hundreds or even thousands of genes can be identified. There are twenty-three pairs of chromosomes in each normal human cell, for a total of forty-six chromosomes in the body. In genetics, a trait is an attribute that has been discovered through a gene, and it is often governed by a huge amount of genes. Certain characteristics are produced by mutated genes that have been handed down through the generations or that have arisen as a result of a novel gene mutation (Wang, 2002; Uhlmann, 2016).

When it comes to types of substances found in the human body, proteins are by far the most plentiful. Proteins are greater than just primary components for connective tissues, muscles, skin, and other organs; they also function as important signaling molecules (Cortés et al., 2003; Ichijima et al., 2012). They're also necessary for the creation of enzymes, as previously stated. When it comes to the body's chemical processes and activities, enzymes are big proteins that regulate and carry out virtually all of them. The human body generates thousands of various enzymes that are used in various processes. So the kinds and amounts of proteins generated by the body determine the overall function and structure of that body. Genes, which have been discovered on chromosomes, are responsible for regulating protein creation (Mills et al., 2003; Greenberg & Bourc'his, 2019).

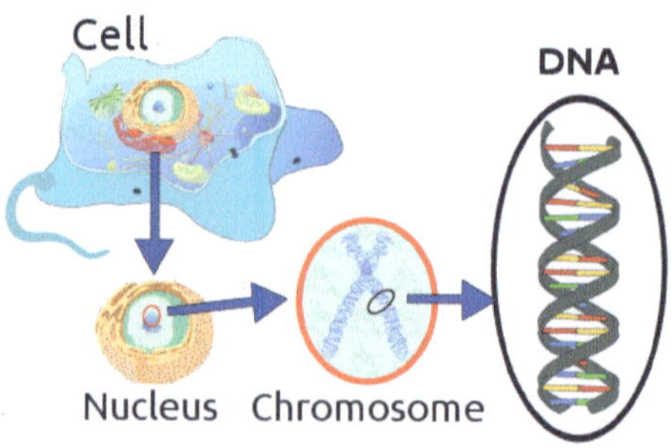


**Figure 5.1.** Cellular Chromosomes and DNA.

**Source:** https://www.microscopemaster.com/images/512px-Eukaryote_DNA.png

A person's genotype (or genome) is a collection of genes or genetic makeup that is unique to them. This shows in the entire collection of commands for how an individual's body creates proteins and, as a result, how that individual's body should be generated and function (Eden & Cedar, 1994; Pollack et al., 2002).

The phenotype of a person is comprised of the physical structural and functional components of that individual. The phenotype is how a person's genotype manifests itself; however, not all of the genotype's orders will be carried out. It is possible that the environment (particularly illnesses and nutrition) and other variables, some of which are unknown, play a role in deciding whether or not a gene is expressed. A karyotype is a depiction of a person's complete collection of chromosomes in their cells, which is called a chromosomal map (Futscher et al., 2002; Raymond et al., 2009).

## 5.2. CHARACTERISTICS OF DNA

Genes are made up of deoxyribonucleic acid. It is DNA that contains the coding and blueprints that are used in the production of proteins. The size of the gene changes depending on the size of the proteins that it encodes for. Every DNA molecule is a long double-helical molecule with millions of steps, comparable to a spiral staircase in appearance. Four distinct types of molecules referred to as bases, are paired together to form the steps of the stairway (nucleotides). During every step of the cycle, the base adenine (A) is linked with the base thymine (T), and in every phase of the cycle, the base guanine (G) is coupled with the base cytosine (C). Every enormously long DNA molecule found within one of the chromosomes has become entangled with one another (Wang, 1985; Oh et al., 2002).

### 5.2.1. DNA Structure

DNA is the genetic substance of the cell, and it has been found in chromosomes in the mitochondria and nucleus of the cell.

The nucleus of the cell, except for some cells (such as egg, sperm and red blood cells), includes twenty-three pairs of chromosomes. A chromosome carries a large number of genes. A gene is a piece of DNA that contains the instructions for making a protein (Lauer et al., 1980; El Hassan, & Calladine, 1997).

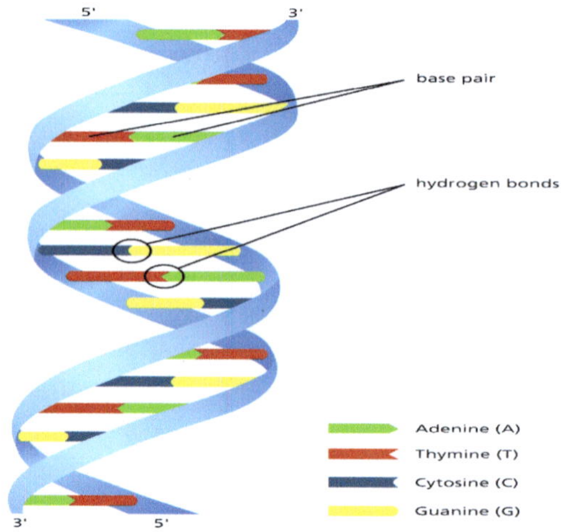


**Figure 5.2.** The double helix structure of DNA is depicted in this picture.

**Source:** https://www.yourgenome.org/facts/what-is-dna

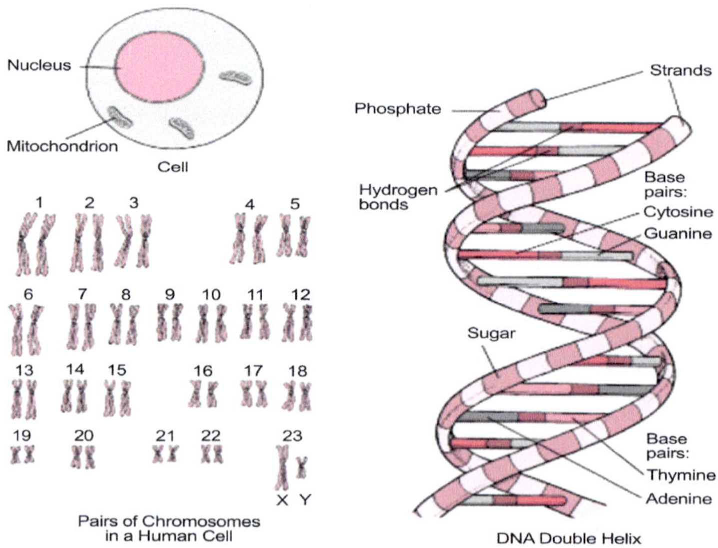


**Figure 5.3.** DNA Structure in Detail.

**Source:** https://www.msdmanuals.com/home/fundamentals/genetics/genes-and-chromosomes

DNA is a helical staircase-like elongated, twisted double helix. Double strands of sugar (deoxyribose) and molecules of phosphate are joined by pairings of 4 molecules known as bases that comprise the staircase's steps. Thymine and adenine are paired in the stages, whereas cytosine and guanine are paired. An H bond binds every pair of bases together. A sequence of bases makes up a gene. The 3 -base patterns may represent an amino acid (proteins' building blocks) or other data (Greenwood et al., 1989).

### 5.2.2. Synthesizing Proteins

Proteins are formed by a long chain of amino acids that have been linked together one by one when they are broken down. There are a total of twenty various amino acids that may be utilized in the process of protein production. It is recommended that some nutrients be received from food, while others are created by the body's enzymes. Whenever you place together a chain of amino acids, the amino acids fold in on themselves to produce a sophisticated 3-D structure. The folded structure's ability to operate in the body is determined by the way it is folded. Because the folding of a protein is determined by the amino acid sequence, each pattern results in a unique protein structure. Certain proteins (for example, hemoglobin) contain many folded chains, which is referred to as multi-folding. The DNA molecule carries the instructions for protein production (Fonseca Guerra & Bickelhaupt, 1999; Wiencke et al., 2000).

### 5.2.3. Coding

The order wherein the bases (adenine (A), thymine (T), guanine (G), and cytosine (C)) are placed in DNA encodes data. The code is written with the help of triplets. However, the bases are assembled in groups of three. A certain 3-base DNA pattern codes for specific directives, like adding a single amino acid to a chain of amino acids. Alanine inclusion is noted by the letters GCT, whereas valine inclusion is noted by the letters GTT (Pietramellara et al., 2001; Rezvanfar et al., 2008). Thus, the amino acid pattern inside a protein is determined by the arrangement of triplet base pairs within the gene for that protein on a molecule of DNA, rather than the amino acid sequence itself. A protein is created through the method of transcription and translation, which are two phases in the process of translating genetic data into protein (Fernández-Santos et al., 2007).

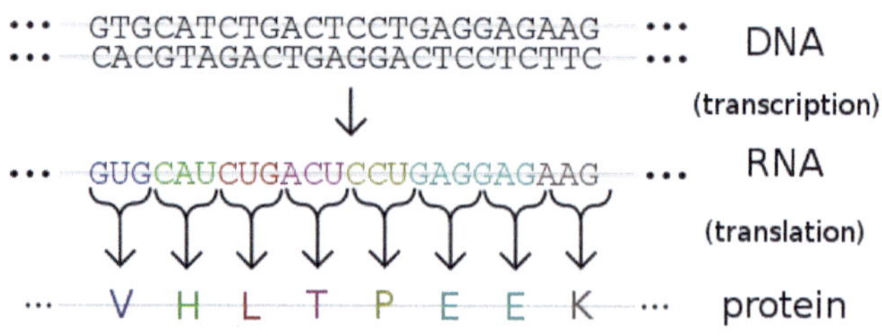


**Figure 5.4.** DNA genetic Coding.

**Source:** https://commons.wikimedia.org/wiki/File:Genetic_code.svg

## 5.2.4. Transcription and Translation

A transcription reaction occurs when information programmed in DNA is transferred to a ribonucleic acid (RNA). The nucleotide uracil (U) replaces the base thymine in RNA, resulting in a long chain of bases that is equivalent to a strand of DNA in length. Consequently, RNA, like DNA, has information encoded in triplets (Sakamoto et al., 1993).

As transcription begins, a segment of DNA's two helices unfolds and uncoils, forming a new strand of DNA. One of the unfolded DNA strands is used as a template for the creation of an RNA strand that corresponds to the unfolded DNA strand. Messenger RNA (mRNA) is the strand of RNA that corresponds to the corresponding strand of RNA. Following its separation from the DNA, mRNA leaves the nucleus and travels to the cytoplasm of the cell. The messenger RNA (mRNA) subsequently creates a connection with a ribosome, a tiny component of the cell that is responsible for protein synthesis (Sugiura & Matsumoto, 1993; Friedman & Nunnari, 2014).

While translation is taking place, the ribosome is given specific instructions about which amino acids to link together in what order. An extremely short kind of RNA known as transfer RNA (tRNA) is responsible for transporting amino acids from the amino acid pool to the ribosome. With each tRNA molecule added to the forming protein chain, which is folded into a complex 3-D shape under the influence of neighboring molecules known as the chaperone molecules, each tRNA molecule adds one amino acid to the growing protein chain (Schardl et al., 1984; Cline, 2012).

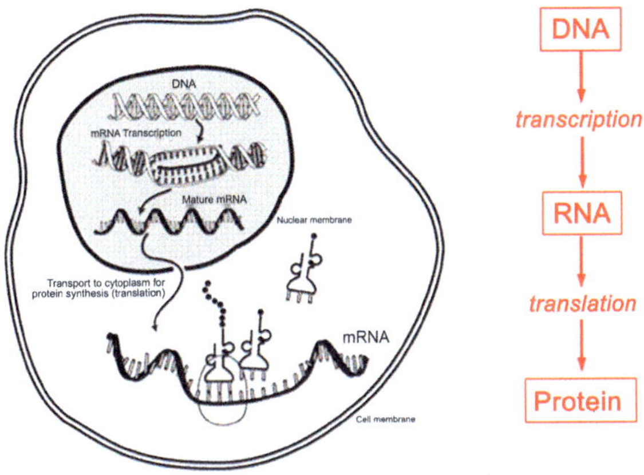


**Figure 5.5.** The translation and transcription of DNA.

**Source:** https://www.ncbi.nlm.nih.gov/Class/MLACourse/Modules/MolBioReview/central_dogma.html

## 5.2.5. Control of Gene Expression

A person's body contains several different kinds of cells, including liver cells, muscle cells, and heart cells. Such cell varies in appearance and behavior, as well as the chemical molecules they generate. Each cell, on the other hand, is a descendent of a single fertilized egg cell and hence has identical DNA. Since various genes are defined in various cells, cells have extremely distinct looks and activities. DNA also contains data when a gene must be expressed. The kind of tissue, the person's age, the existence of certain chemical signals, and a variety of other factors and methods all influence gene expression. Although our understanding of the various variables and processes that affect gene expression is quickly expanding, most of such factors and mechanisms remain unknown (Voigt et al., 2008).

The processes through which genes regulate one another are extremely complex. Chemical markers on genes show wherever transcription must start and stop. Different chemical compounds in and around the DNA block (like histones) allow transcription. In addition, an RNA strand known as antisense RNA may link with a complementary strand of mRNA and prevent translation (Collins & Seville, 1990; Shao et al., 2009).

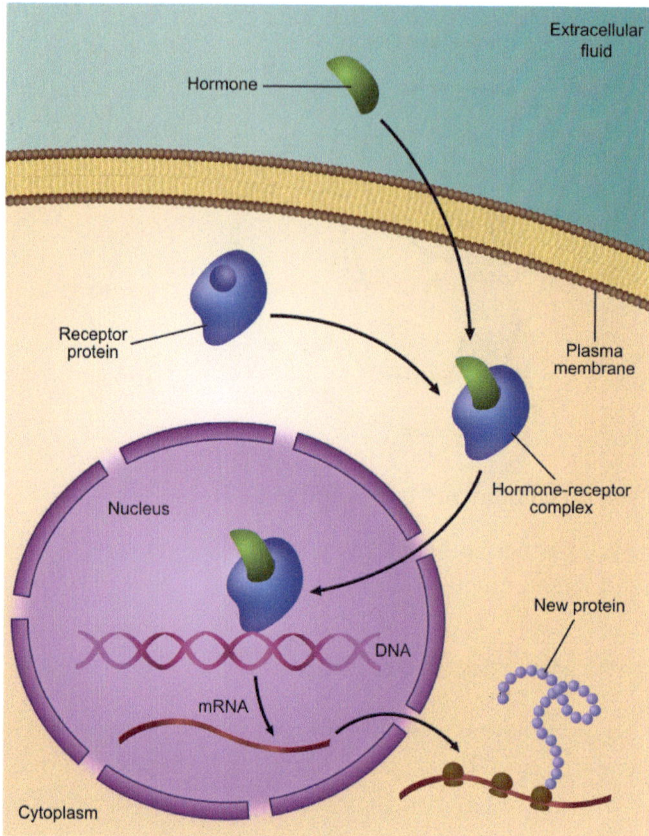


**Figure 5.6.** The control of Gene Expression.

**Source:** https://sphweb.bumc.bu.edu/otlt/MPH-Modules/PH/DNA-Genetics/DNA-Genetics7.html

## 5.2.6. Replication

Cells divide in half to reproduce. The molecules of DNA in the parent cell should recreate themselves throughout cellular division since every daughter cell needs a full set of DNA molecules. Replication works similarly to transcription, with the exception that the complete double-strand DNA molecule unfolds and divides in half. Upon splitting, the bases on every strand bond to complementary bases floating adjacent (Adenine with Thymine, and Guanine with Cytosine). Once this procedure is finished, two similar two-strand molecules of DNA are produced (Fischel-Ghodsian, 2003).

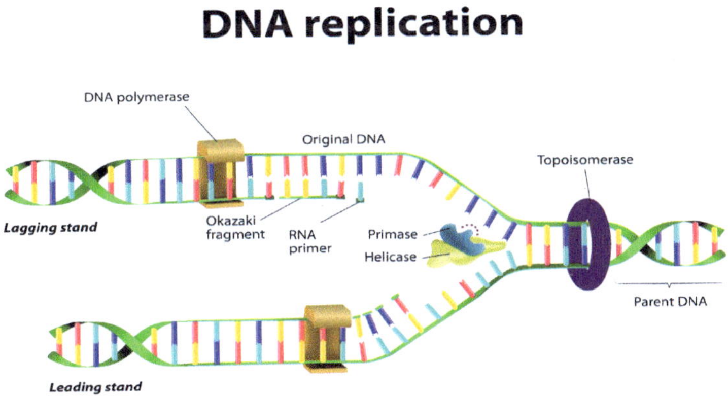


**Figure 5.7.** The DNA replication.

**Source:** https://www.news-medical.net/life-sciences/DNA-Replication-and-Repair.aspx

## 5.2.7. Mutation

Cells include a “proofreading” activity that helps to ensure that bases have been connected correctly when the cell divides and divides again. DNA that has not been duplicated correctly can be repaired using chemical procedures as well as other approaches. However, because of the billions of base pairs involved in the protein manufacturing technique and the complexity of the process, mistakes are conceivable. Such mistakes can be caused by a variety of factors (such as radiation, drugs, or viruses), or they might occur for no apparent reason. Slight differences in the DNA sequence are fairly common and may be detected in practically everyone, even if they are not noticeable. The vast majority of genetic variations do not affect subsequent copies of the gene in a cell. Mutations are mistakes that are repeated in subsequent copies of a file (Gross et al., 1984; Burger et al., 2003).

**Inherited mutations** are disorders that can be passed down from one generation to the next. Mutations that cause harm to reproductive cells may be handed down from generation to generation. Mutations that do not affect reproductive cells cause damage to the ancestors of the mutant cell (for example, cancer), but not to the children of the mutant cell.

Individual or familial characteristics can be distinguished by mutations, and the majority of hazardous mutations are rather infrequent. Polymorphisms are mutations that have become so common that they now affect more than one percent of a population's genetic make-up (for instance, the kinds of human blood O, AB, A, B,). The majority of polymorphisms do not affect the phenotype.

It is possible to mutate little or large sections of DNA. In some cases, the mutation has no effect, while in others it may alter the amino acid sequence of a protein or lower the amount of protein produced, depending on the size and position of the mutation. The amino acid sequence of a protein might vary, resulting in a protein that functions differently or not at all. A protein that is either absent or not functioning properly might be harmful or even fatal. Phenylketonuria, for example, is caused by a mutation in the enzyme phenylalanine hydroxylase, which results in the enzyme being defective or missing. The amino acid phenylalanine (absorbable from food) accumulates in the body as a result of this deficit, resulting in severe intellectual impairment. In rare situations, a mutation results in a beneficial modification. When a person inherits 2 copies of a defective gene, such as the gene of sickle cell, he or she would have sickle cell disease. Whenever a person inherits just 1 copy of the sickle cell gene, he or she acquires limited malaria resistance (an infection of the blood). While malaria resistance may assist a carrier's life, sickle cell disease (which occurs when a person carries 2 copies of the gene) creates signs and consequences that may limit a person's life span.

Natural selection is the idea that mutations that make it harder to survive in a particular environment are less likely to be passed on to children (and hence become less frequent in the population), while mutations that make it easier to survive become more common. As a result, advantageous mutations grow more widespread over time, despite their rarity at first. Evolution is the collective term for the gradual changes that take place with time in an interbreeding population due to mutations and natural selection (Wu et al., 2015).

## 5.3. CHARACTERISTICS OF CHROMOSOMES

A chromosome is a long strand of DNA that contains a significant number of genes and is found in all living organisms. A precise order has been established for the genes on each chromosome, and each gene has been assigned a specific location on the chromosome (known as its locus). The chromosomes include a variety of chemical substances that, in addition to DNA, influence gene activation (Malyarchuk et al., 2004).

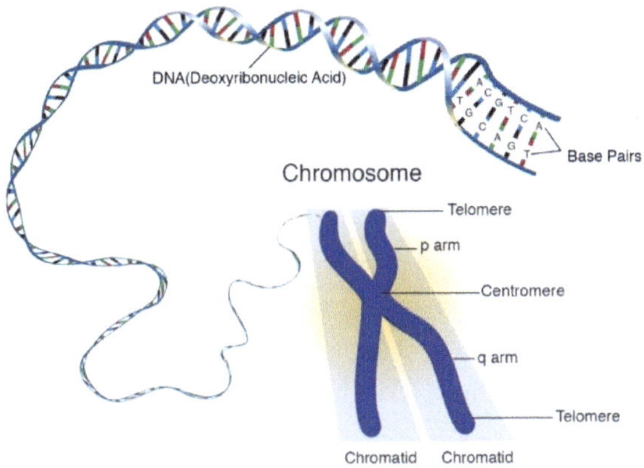


**Figure 5.8.** Chromosomes Structure.

**Source:** https://www.nature.com/articles/ncomms1218

## 5.3.1. Pairing

The nucleus of the human cell includes twenty-three chromosome pairs, for a maximum of forty-six chromosomes, excluding certain cells (such as egg and sperm cells or red blood cells). Usually, every pair has one father's chromosome and one mother's chromosome.

There have been one pair of sex chromosomes and twenty-two pairs of nonsex chromosomes in the human population so far. Pairs of autosomal chromosomes are identical in terms of gene content, size, shape, position, and number for all practical purposes. In addition, there is a backup for the genes on such chromosomes since every member of a pair of autosomal chromosomes contains one copy of each gene that matches the backup copy (Levy et al., 1991).

The sex chromosomes are the 23rd pair of chromosomes (X and Y).

## 5.3.2. Sex Chromosomes

The sex chromosome pair determines whether or not a fetus will grow into a female or a male. Males have two chromosomes, one Y chromosome and one X chromosome. When it comes to men, the X in their DNA is inherited from their mother, and when it comes to women, it is inherited from their father. Females have two X chromosomes, one of which they acquired from their mother and the other which they received from their father. In several ways, sex chromosomes and autosomal chromosomes function in different ways than one another.

The Y chromosome, which is shorter than the other chromosomes, includes the genes that determine whether a person is a male or female, as well as a few extra genes of unknown function. Several of the genes on the X chromosome are involved in functions other than sex differentiation, and as a result, they do not have equivalents on the Y chromosome. Since there has been no 2nd X chromosome in males, such additional X chromosome genes are not paired and almost all of them have been defined. X-linked, genes are those found on the X chromosome (André & Walbot, 1995).

In the autosomal chromosomes, the genes across both sets of chromosomes are ready for full expression. In females, a procedure known as X inactivation turns off many of the genes on one of the 2 X chromosomes (rather than in the eggs present in the ovaries). The fetus's X gene is inactivated initially in its existence.

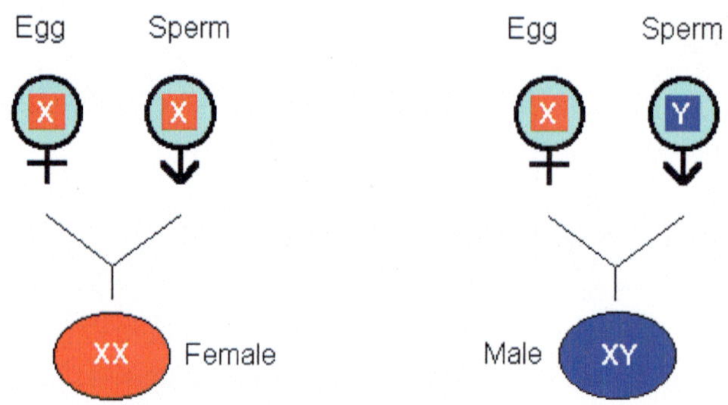


**Figure 5.9.** The Chromosomes of Sex.

**Source:** https://www.web-books.com/MoBio/Free/Ch1C4.htm

The X from the father remains inactive in certain cells, while the X from the mother remains inactive in others. As a result, one cell can include a gene from the person's mother, while another contains a gene from the person's father. The lack of one X chromosome generally leads to mild defects due to X inactivation (like Turner syndrome). As a result, having an X chromosome missing is significantly less detrimental than having an autosomal chromosome lacking (Jorde et al., 2000; Wu, & Sloan, 2019).

### 5.3.3. Inactive X Chromosome

The additional chromosomes in a girl with a condition wherein she contains greater than 2 X chromosomes are usually inactive. As a result, possessing an additional X chromosome creates considerably fewer development problems than possessing an additional nonsex chromosome. Women possessing 3 X chromosomes, for instance, are frequently physically and cognitively healthy. Males with many Y chromosomes may experience mental and physical problems (Kopylov et al., 2018)

### 5.3.4. Chromosome Abnormalities

Chromosome abnormalities can be in a variety of forms. A person's chromosomal count can be aberrant, or abnormal sections on single or many chromosomes can exist. Several of these problems may be detected before delivery.

Serious abnormalities are frequently caused by an abnormal number of nonsex chromosomes. Getting an additional nonsex chromosome, for instance, might be harmful to a fetus or cause anomalies like Down syndrome, which is caused by comprising 3 copies of chromosome-21. The fetus will die if it lacks a nonsex chromosome.

Large sections of a chromosome can be aberrant, generally because a full piece had been deleted or accidentally inserted into another chromosome (known as translocation). Translocation of the region of chromosome-9 onto chromosome-22 can, for instance, induce chronic myelogenous leukemia. Such a condition may be inherited or caused by a novel mutation (Ricchetti et al., 1999).

### 5.3.5. Mitochondrial Chromosomes

Mitochondria are small structures found within the cells that manufacture energy-producing chemicals. Every mitochondrion, rather than other structures of a cell, has its complete set of chromosomes. Such chromosome includes mitochondrial DNA that encodes among some, though not all, of the mitochondrion's proteins. Mitochondrial DNA is normally solely inherited from one's mother since, whenever an egg has been fertilized, just the egg's mitochondria constitute part of the growing embryo. Mitochondria from sperm don't normally make it into the growing fetus (Marande & Burger, 2007).

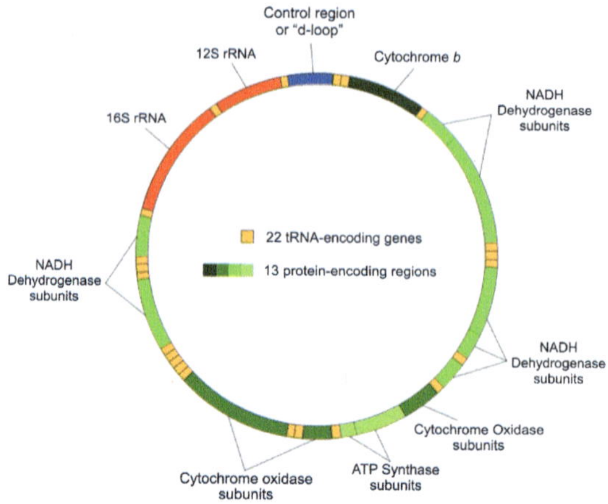


**Figure 5.10.** The mitochondrial chromosomes are represented graphically.

**Source:** http://www.contexo.info/dna_basics/chromosomes/mitochondria/

## 5.3.6. Traits

Every gene-determined property is referred to as a trait. Several features are influenced by the interactions of multiple genes. Multiple genes, such as those regulating development, hunger, muscular mass, and level of activity, are likely to affect a person's height. However, the function of one gene determines the expression of several features.

Certain differences in characteristics, like the color of the eye or blood group, are deemed acceptable. Albinism, Marfan syndrome, and Huntington's disease are examples of changes that affect bodily function or structure and are classified as diseases. Furthermore, not every gene anomaly is damaging in the same way. A single copy of the sickle cell gene, for instance, may defend you from malaria, whereas 2 copies may induce sickle cell anemia (Nosek et al., 2006).

## 5.3.7. Genetic Disorders

A faulty gene causes a genetic disease, which is a negative characteristic. It's possible that the defective gene is inherited or that it develops naturally as a consequence of a novel mutation. Anomalies of the genes are quite prevalent. Each person possesses one hundred to four hundred defective genes on average (it varies from person to person). The matching gene on the opposite chromosome in the pair, on either hand, is usually normal and avoids any negative consequences. The likelihood of a person containing 2 copies of the same defective gene in the general public is extremely low. Kids who have been the offspring of closest blood relatives, on the other hand, have a better likelihood. Children of parents who married inside an isolated group, like the Amish or Mennonites, have a greater chance (Corach et al., 2010).

The nucleosome is a kind of DNA molecule. The nucleosome is made up of a DNA double helix (*grey*) folded around a central particle of histone proteins. Along with the chromosomal DNA, nucleosomes are typically two hundred nucleotide pairs apart.

Cells' capacity to retain, access, and interpret the genetic information necessary to create and sustain a live creature is critical to life. At cellular division, genetic data is communicated to a cell to its daughter cells, as well as from the first generation of an organism to the second through the sexual cells of the organism. Such commands are stored as genes in each living cell, data that contains elements that define the features of a species as a whole as well as the people within it.

The molecular structure of genes piqued scientists' interest as soon as genetics became a discipline around the turn of the 20th century. During the life of a multicellular creature, the data in genes are replicated and transferred hundreds of thousands of times from parent cell to daughter cell, and it survives the procedure virtually intact. What kind of molecule might be capable of this precise and nearly limitless reproduction while also directing the growth of an organism and the day-to-day existence of a cell? What type of commands may be found in genetic data? How are these commands physically structured to fit the huge volume of data necessary for the growth and sustenance of even the smallest creature into the little area of a cell?

The discovery of DNA as the probable carrier of genetic data was another significant breakthrough in the 1940s. However, the technique by which genetic data is reproduced and transmitted from one cell to another cell, as well as how proteins are dictated by DNA commands, remained a mystery. When James Watson and Francis Crick discovered the DNA structure in 1953, the enigma was suddenly resolved. The DNA structure resolved the dilemma of how data in this molecule could be reproduced or replicated almost instantaneously. It also offered the 1st indications of how a DNA molecule may encode protein-making commands. The concept that DNA is the genetic information is so essential to biological knowledge today that it's hard to comprehend what a huge conceptual void this finding bridged (Tambets et al., 2004).

Genes are transported on chromosomes, which had been found in the 19th century as microfilaments structures in the nucleus of a eukaryotic cell which become apparent when the cell divides, long before biologists knew the DNA structure. Chromosomes were later discovered to contain both DNA and protein once biochemical analysis became practicable. We currently know that DNA contains the cell's genetic data. The protein elements of chromosomes, on the other hand, serve primarily to bundle and manage the immensely long DNA molecules so that they can fit within cells and be readily available.

DNA is shown to be the genetic information in an experiment. These tests, conducted in the 1940s, demonstrated that introducing pure DNA to a bacterium altered its characteristics and that this change was reliably handed down to later generations (Falk et al., 2016).

## REFERENCES

1. André, C. P., & Walbot, V. (1995). Pulsed-field gel mapping of maize mitochondrial chromosomes. *Molecular and General Genetics MGG, 247*(2), 255-263.
2. Burger, G., Forget, L., Zhu, Y., Gray, M. W., & Lang, B. F. (2003). Unique mitochondrial genome architecture in unicellular relatives of animals. *Proceedings of the National Academy of Sciences, 100*(3), 892-897.
3. Cline, S. D. (2012). Mitochondrial DNA damage and its consequences for mitochondrial gene expression. *Biochimica et Biophysica Acta (BBA)-Gene Regulatory Mechanisms, 1819*(9-10), 979-991.
4. Collins, R. A., & Seville, B. J. (1990). Independent transfer of mitochondrial chromosomes and plasmids during unstable vegetative fusion in Neurospora. *Nature, 345*(6271), 177-179.
5. Corach, D., Lao, O., Bobillo, C., Van Der Gaag, K., Zuniga, S., Vermeulen, M., ... & Kayser, M. (2010). Inferring continental ancestry of Argentineans from autosomal, Y-chromosomal and mitochondrial DNA. *Annals of human genetics, 74*(1), 65-76.
6. Cortés, F., Pastor, N., Mateos, S., & Domınguez, I. (2003). Roles of DNA topoisomerases in chromosome segregation and mitosis. *Mutation Research/Reviews in Mutation Research, 543*(1), 59-66.
7. Eden, S., & Cedar, H. (1994). Role of DNA methylation in the regulation of transcription. *Current opinion in genetics & development, 4*(2), 255-259.
8. El Hassan, M. A., & Calladine, C. R. (1997). Conformational characteristics of DNA: empirical classifications and a hypothesis for the conformational behaviour of dinucleotide steps. *Philosophical Transactions of the Royal Society of London. Series A: Mathematical, Physical and Engineering Sciences, 355*(1722), 43-100.
9. Falk, M. J., Decherney, A., & Kahn, J. P. (2016). Mitochondrial replacement techniques—implications for the clinical community. *The New England journal of medicine, 374*(12), 1103.
10. Fernández-Santos, M. R., Martínez-Pastor, F., García-Macías, V., Esteso, M. C., Soler, A. J., Paz, P., ... & Garde, J. J. (2007). Sperm characteristics and DNA integrity of Iberian red deer (Cervus elaphus hispanicus) epididymal spermatozoa frozen in the presence of enzymatic and nonenzymatic antioxidants. *Journal of andrology, 28*(2), 294-305.

11. Fischel-Ghodsian, N. (2003). Mitochondrial deafness. *Ear and hearing*, *24*(4), 303-313.
12. Fonseca Guerra, C., & Bickelhaupt, F. M. (1999). Charge transfer and environment effects responsible for characteristics of DNA base pairing. *Angewandte Chemie International Edition*, *38*(19), 2942-2945.
13. Friedman, J. R., & Nunnari, J. (2014). Mitochondrial form and function. *Nature*, *505*(7483), 335-343.
14. Futscher, B. W., Oshiro, M. M., Wozniak, R. J., Holtan, N., Hanigan, C. L., Duan, H., & Domann, F. E. (2002). Role for DNA methylation in the control of cell type–specific maspin expression. *Nature genetics*, *31*(2), 175-179.
15. Greenberg, M. V., & Bourc'his, D. (2019). The diverse roles of DNA methylation in mammalian development and disease. *Nature reviews Molecular cell biology*, *20*(10), 590-607.
16. Greenwood, M. S., Hopper, C. A., & Hutchison, K. W. (1989). Maturation in larch: I. Effect of age on shoot growth, foliar characteristics, and DNA methylation. *Plant Physiology*, *90*(2), 406-412.
17. Gross, S. R., Hsieh, T. S., & Levine, P. H. (1984). Intramolecular recombination as a source of mitochondrial chromosome heteromorphism in Neurospora. *Cell*, *38*(1), 233-239.
18. Ichijima, Y., Sin, H. S., & Namekawa, S. H. (2012). Sex chromosome inactivation in germ cells: emerging roles of DNA damage response pathways. *Cellular and Molecular Life Sciences*, *69*(15), 2559-2572.
19. Jorde, L. B., Watkins, W. S., Bamshad, M. J., Dixon, M. E., Ricker, C. E., Seielstad, M. T., & Batzer, M. A. (2000). The distribution of human genetic diversity: a comparison of mitochondrial, autosomal, and Y-chromosome data. *The American Journal of Human Genetics*, *66*(3), 979-988.
20. Kopylov, A. T., Ponomarenko, E. A., Ilgisonis, E. V., Pyatnitskiy, M. A., Lisitsa, A. V., Poverennaya, E. V., ... & Archakov, A. I. (2018). 200+ protein concentrations in healthy human blood plasma: targeted quantitative SRM SIS screening of Chromosomes 18, 13, Y, and the mitochondrial chromosome encoded proteome. *Journal of proteome research*, *18*(1), 120-129.
21. Lauer, E., Helming, C., & Kandler, O. (1980). Heterogeneity of the species Lactobacillus acidophilus (Moro) Hansen and Moquot as revealed by

biochemical characteristics and DNA-DNA hybridisation. *Zentralblatt für Bakteriologie: I. Abt. Originale C: Allgemeine, angewandte und ökologische Mikrobiologie*, *1*(2), 150-168.

22. Levy, A. A., Andre, C. P., & Walbot, V. (1991). Analysis of a 120-kilobase mitochondrial chromosome in maize. *Genetics*, *128*(2), 417-424.
23. Malyarchuk, B., Derenko, M., Grzybowski, T., Lunkina, A., Czarny, J., Rychkov, S., ... & Miścicka-Śliwka, D. (2004). Differentiation of mitochondrial DNA and Y chromosomes in Russian populations. *Human biology*, 877-900.
24. Marande, W., & Burger, G. (2007). Mitochondrial DNA as a genomic jigsaw puzzle. *Science*, *318*(5849), 415-415.
25. Mills, K. D., Ferguson, D. O., & Alt, F. W. (2003). The role of DNA breaks in genomic instability and tumorigenesis. *Immunological reviews*, *194*(1), 77-95.
26. Nosek, J., Tomaska, L., Bolotin-Fukuhara, M., & Miyakawa, I. (2006). Mitochondrial chromosome structure: an insight from analysis of complete yeast genomes. *FEMS yeast research*, *6*(3), 356-370.
27. Oh, S. J., Cho, S. J., Kim, C. O., & Park, J. W. (2002). Characteristics of DNA microarrays fabricated on various aminosilane layers. *Langmuir*, *18*(5), 1764-1769.
28. Pietramellara, G., Franchi, M., Gallori, E., & Nannipieri, P. (2001). Effect of molecular characteristics of DNA on its adsorption and binding on homoionic montmorillonite and kaolinite. *Biology and Fertility of Soils*, *33*(5), 402-409.
29. Pollack, J. R., Sørlie, T., Perou, C. M., Rees, C. A., Jeffrey, S. S., Lonning, P. E., ... & Brown, P. O. (2002). Microarray analysis reveals a major direct role of DNA copy number alteration in the transcriptional program of human breast tumors. *Proceedings of the National Academy of Sciences*, *99*(20), 12963-12968.
30. Raymond, J. J., van Oorschot, R. A., Gunn, P. R., Walsh, S. J., & Roux, C. (2009). Trace evidence characteristics of DNA: a preliminary investigation of the persistence of DNA at crime scenes. *Forensic Science International: Genetics*, *4*(1), 26-33.
31. Rezvanfar, M. A., Sadrkhanlou, R. A., Ahmadi, A., Shojaei-Sadee, H., Rezvanfar, M. A., Mohammadirad, A., ... & Abdollahi, M. (2008). Protection of cyclophosphamide-induced toxicity in reproductive tract

histology, sperm characteristics, and DNA damage by an herbal source; evidence for role of free-radical toxic stress. *Human & experimental toxicology*, *27*(12), 901-910.

32. Ricchetti, M., Fairhead, C., & Dujon, B. (1999). Mitochondrial DNA repairs double-strand breaks in yeast chromosomes. *Nature*, *402*(6757), 96-100.
33. Sakamoto, T., Shirai, M., Asayama, M., Aida, T., Sato, A., Tanaka, K., ... & Nakano, M. (1993). Characteristics of DNA and multiple rpoD homologs of Microcystis (Synechocystis) strains. *International Journal of Systematic and Evolutionary Microbiology*, *43*(4), 844-847.
34. Schardl, C. L., Lonsdale, D. M., Pring, D. R., & Rose, K. R. (1984). Linearization of maize mitochondrial chromosomes by recombination with linear episomes. *Nature*, *310*(5975), 292-296.
35. Shao, R., Kirkness, E. F., & Barker, S. C. (2009). The single mitochondrial chromosome typical of animals has evolved into 18 minichromosomes in the human body louse, Pediculus humanus. *Genome Research*, *19*(5), 904-912.
36. Sugiura, Y., & Matsumoto, T. (1993). Some characteristics of DNA strand scission by macromolecular antitumor antibiotic C-1027 containing a novel enediyne chromophore. *Biochemistry*, *32*(21), 5548-5553.
37. Tambets, K., Rootsi, S., Kivisild, T., Help, H., Serk, P., Loogväli, E. L., ... & Villems, R. (2004). The western and eastern roots of the Saami—the story of genetic "outliers" told by mitochondrial DNA and Y chromosomes. *The American Journal of Human Genetics*, *74*(4), 661-682.
38. Uhlmann, F. (2016). SMC complexes: from DNA to chromosomes. *Nature reviews Molecular cell biology*, *17*(7), 399-412.
39. Voigt, O., Erpenbeck, D., & Wörheide, G. (2008). A fragmented metazoan organellar genome: the two mitochondrial chromosomes of *Hydra magnipapillata*. *BMC genomics*, *9*(1), 1-10.
40. Wang, J. C. (1985). DNA topoisomerases. *Annual review of biochemistry*, *54*(1), 665-697.
41. Wang, J. C. (2002). Cellular roles of DNA topoisomerases: a molecular perspective. *Nature reviews Molecular cell biology*, *3*(6), 430-440.
42. Wiencke, C., Gómez, I., Pakker, H., Flores-Moya, A., Altamirano, M., Hanelt, D., ... & Figueroa, F. L. (2000). Impact of UV-radiation

on viability, photosynthetic characteristics and DNA of brown algal zoospores: implications for depth zonation. *Marine Ecology Progress Series*, *197*, 217-229.

43. Wu, Z., & Sloan, D. B. (2019). Recombination and intraspecific polymorphism for the presence and absence of entire chromosomes in mitochondrial genomes. *Heredity*, *122*(5), 647-659.
44. Wu, Z., Cuthbert, J. M., Taylor, D. R., & Sloan, D. B. (2015). The massive mitochondrial genome of the angiosperm Silene noctiflora is evolving by gain or loss of entire chromosomes. *Proceedings of the National Academy of Sciences*, *112*(33), 10185-10191.

CHAPTER 6

# CATALYSIS AND THE USE OF ENERGY BY CELLS

CONTENTS

## 6.1. INTRODUCTION

Above everything, one quality of organisms distinguishes them from non-living things almost magically: they establish as well as preserve order in a world that is continually going toward more chaos. To establish this equilibrium, the cells of a live individual should carry out an infinite number of chemical processes (Alberts et al., 2002). Several of these responses involve the disassembly or modification of biological molecules enzymes, sugars, nucleotides, and lipids—to provide the cell with the numerous molecules it needs. In other processes, such tiny molecules are utilized to synthesize an incredibly diversified array of proteins, nucleic acids, as well as other macromolecules which provide organisms with their most distinguishing characteristics. Every cell may be thought of as a little chemical plant, capable of completing hundreds of events per second (Babcock & Wikström, 1992).

Usually, the chemical processes carried out by a cell could exist mostly at temperatures far greater than any of those found within cells. As just a result, every reaction necessitates a unique increase in chemical reactivity. This need is critical since this enables the cell for regulating every response. Management is achieved via specialized proteins known as enzymes, each one enhances or catalyzes, one of the numerous reactions that such a given molecule may undertake. Typically, enzyme-catalyzed responses are performed in succession, with a result of a single reaction serving as the raw products, or substrates, for the subsequent reaction (Centi et al., 2013). These lengthy sequential reaction paths are then linked to make a maze of interrelated reactions which allow the cell to stay alive, raise, and procreate.

In a classic cell, several metabolic paths, as well as their interrelations, are shown. Around 600 important metabolic responses are schematically depicted, with every particle in a metabolism denoted by a filled circular pattern (Centi & Perathoner, 2009).

In cells, two contradictory channels of chemical reactions occur: (1) catabolic reactions decompose food products into simpler compounds, providing the cell with potential power along with some of the organic compounds it requires as basic components; and (2) biosynthetic or anabolic, paths are using the energy generated by catabolism to push formation of numerous other molecules as it makes up the cell. Together, these two pairs of responses comprise the cell's metabolism (Armor, 1999).

Numerous elements of cell metabolism are covered in the conventional field of biochemistry and do not have to be discussed here. However, the fundamental principles governing how cells get energy through their surroundings and employ it to generate order are critical to cell biology. We start by discussing why living creatures require a steady supply of energy (Bard, 1979).

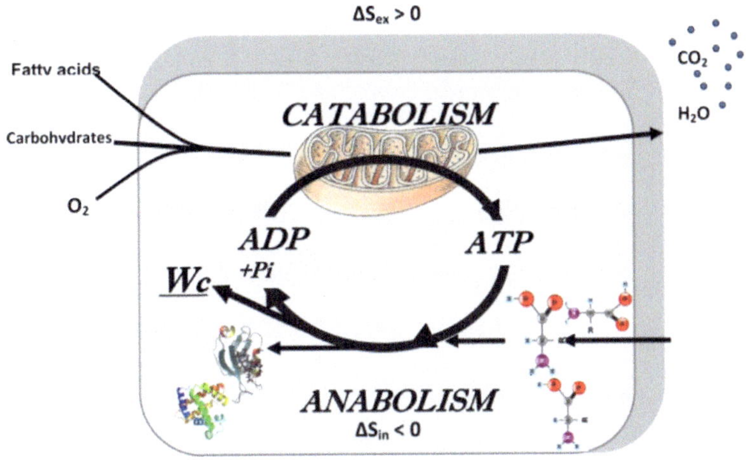


**Figure 6.1.** General scheme of cellular metabolism.

**Source:** researchgate.net/figure/General-scheme-of-cellular-metabolism-A-cell-is-a-thermodynamically-open-system-in_fig3_24366278

## 6.2. BIOLOGICAL ORDER IN CELLS

The worldwide potential for items towards becoming disturbed is displayed in basic scientific law, the 2nd law of thermodynamics, which states that its degree of irregularity can only raise in the world, or some other closed system (a gathering of the matter entirely separated from the remaining universe). This law has such far-reaching ramifications for all living creatures that it ought to be rephrased numerous times (Sato et al., 1991).

For instance, we may express the second rule as a probabilistic law, stating that systems will move freely to the configurations with the highest likelihood. If we examine a box of 100 coins, all of which are laying heads up, a sequence of mishaps will cause the configuration to shift to a combination of 50 tails and 50 heads. The cause for this is easy: there are several different ways to organize the specific coins in the combination to accomplish the 50-50 result, but just one way of keeping all the coins directed heads up (Marquez-Curtis et al., 2015). We describe the 50-50 blend as more "disordered" since it is the most likely. For a similar cause, this is a frequent thing for someone's living space for becoming progressively messy without conscious effort: disorganized is a natural method that demands conscious effort to reverse.

A system's degree of the disorder may be measured. The entropy of the system is the amount we are using to quantify this disorder: the higher the unalignment, the higher the entropy. As a result, a third method of describing the 2nd law of thermodynamics is suggested as systems would alter instantaneously to entropy-rich provisions (Brownstein & Tarr, 1979).

Living cells generate order by staying alive, growing, and forming multicellular life, which may show up to disobey the 2nd law of thermodynamics. What gives that this possible? The response is whether a cell is not an individual entity: it absorbs the energy out of its surroundings by the type of food, photons from the sun (or indeed inorganic molecules on its own, and that in few chemosynthetic bacteria), and utilizes it to create order inside itself (Kim et al., 2009). The portion of the energy used by the cell is transformed into heat during the redox process that creates order. The energy is released into the cell's surroundings, causing it to become disordered, causing the overall entropy—that of a cell and its surroundings—to rise, as the laws of science demand.

Consider a cell resting in a sea of stuff that represents the remaining universe to grasp the rules regulating these energy transformations (Asami et al., 1980). The cell generates inner order as it gets to live and grows. However, when it synthesizes molecules as well as combines them into cell structures, it emits heat energy. Heat is perhaps the most disorganized kind of energy, consisting of the random bumping of molecules. Whenever a cell emits heat into the sea, it enhances the impact of molecular movement (thermal movement), which enhances the sea's variation or disorder. Since the rise in order within the cell is mostly offset by a bigger reduction (gain in entropy) in the neighboring sea of matter, the second law of thermodynamics is met (Beyeler et al., 2007).

What is the source of the heat which the cell produces? An essential rule of thermodynamics is encountered here. Energy may be changed from one form to another, but it could not be generated or annihilated, according to the basic rule of thermodynamics. The quantity of energy in various types will vary like the consequence of chemical processes within the cell, nonetheless, the primary rule states that the overall quantity of energy must remain constant (Hahn et al., 1995). A mammalian cell, for instance, consumes food and transfers part of the energy contained in chemical bonds among the atoms of such carbohydrates and fats (chemical reactions energy) into the random thermal movement of molecules (heat energy). If the processes within the cell are to lead the world as one to form more disorganized, this transformation of molecular energy into heat energy is necessary by the second law (Baulig et al., 2003).

Except if the thermal reactions within the cell are closely connected to the mechanisms that build molecular structure, the cell could not fit from the latent heat it emits. The metabolism of a cell is distinguished from the inefficient fuel combustion in a fire by the close linkage of heat generation to a rise in the organization. We'll show how this pairing occurs later in this section (Baulig et al., 2003). For the time being, it's enough to acknowledge that an allowed relationship between the "burning" of diet molecules and the production of biological order is essential if cells are to construct as well as sustain an island of order in an otherwise chaotic environment (Loverdo et al., 2008).

## 6.3. PHOTOSYNTHESIS

All creatures survive by using the energy available in the chemical bonds of organic compounds created by other creatures. Additionally, the compounds in food offer the elements required for species to create new living matter. Certain animals receive their sustenance via the consumption of other creatures. However, creatures that consume plants are in the lower region of the animal food supply chain. Plants, in return, capture energy straight from the sun. As a consequence, all energy consumed by animal cells is eventually taken from the sun (Farquhar et al., 2001).

Photosynthesis in a plant as well as photosynthetic microorganisms is how solar light penetrates the world. Photosynthesis is the process by which electromagnetic energy from the sun is turned into chemical reaction energy inside the cell. Plants may receive all of the atoms they require from non-living matter (Gust & Moore, 1989): carbon through distinctive carbon dioxide, oxygen as well as hydrogen from fluids, nitrogen from ammonia and nitrates in the soil, and additional components in trace quantities from artificial salts in the soil. They convert such atoms to sugars, amino acids, nucleotides, and fatty acids using the energy they get from sunshine. These tiny molecules are then transformed into proteins, nucleic acids, polysaccharides, and lipids, which comprise the plant. Every one of such compounds acts as simple sugars for animals if the plants are subsequently consumed (Evans, 2013).

Photosynthesis is a 2-phase procedure. The first step involves the collection and brief storage of solar power as chemical reaction energy in specialized tiny molecules that operate as energy carriers and chemically reactive units. As a waste material of this initial step, molecular oxygen (O2 gas) is produced through the dissociation of water by sunlight (Eberhard et al., 2008).

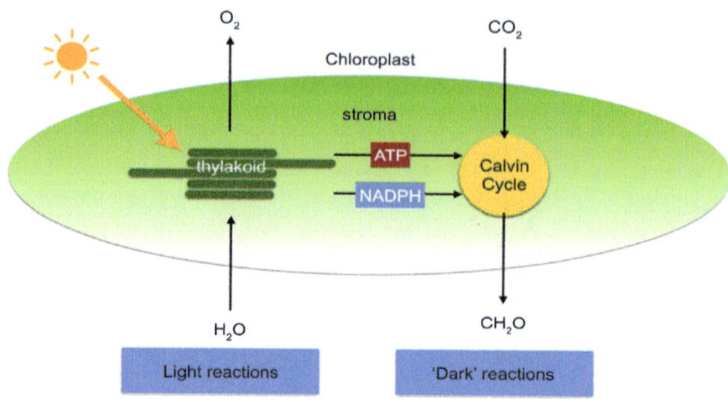


**Figure 6.2.** The two stages of photosynthesis.

**Source:** Johnson, Matthew P. "Photosynthesis." Essays in biochemistry *vol. 60*,3 (2016): 255-273. doi:10.1042/EBC20160016

In the second phase, the energy-carrying molecules are being used to assist in the carbon fixation procedure, which produces sugars from carbon dioxide gas ($CO_2$) and water ($H_2O$), thus also offering a beneficial origin of deposited chemical bond energy as well as materials both for the plant as well as any animals that consume it. The exquisite mechanics behind these two phases of photosynthesis are described in detail (Ke, 2001).

The overall outcome of the whole photosynthetic process, in terms of the green plant, may be stated simply by the equation.

$$\text{light energy} + CO_2 + H_2O \rightarrow \text{sugars} + O_2 + \text{heat energy}$$

Sugars are subsequently employed as a form of chemical reaction energy and also as a supply of resources for the production of the many other tiny and big organic molecules required by the plant cell (Hohmann-Marriott et al., 2011).

## 6.4. OXIDATION OF ORGANIC MOLECULES

It is either sugars whose, such a plant has photosynthesized as food for their own or a combination of big as well as tiny particles which an animal has consumed, all plant and animal cells are driven by energy deposited in the chemical bonds of molecules. Living things should collect this energy in a usable state to utilize it to live, develop, and propagate. Energy is taken from simple sugars by a procedure of slow oxidation, or regulated burning, in both flora and fauna (Sariaslani et al., 1989).

The earth's environment includes a lot of oxygen, and in the existence of oxygen, the most metabolically stable phase of carbon is $CO_2$, while the most metabolically stabilized type of hydrogen is H2O. A cell may get energy from sugars or even other organic matter by combining their hydrogen and carbon along with oxygen to form $CO_2$ and $H_2O$, correspondingly, in a procedure termed respiration (Wang & Stahl, 2020).

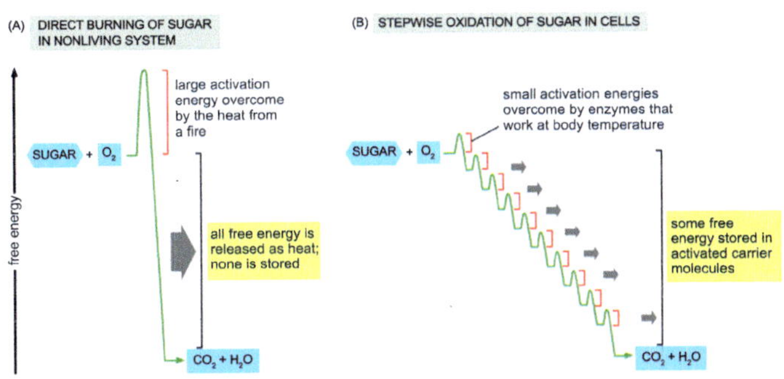


**Figure 6.3.** Graphic picture of the measured stepwise oxidation of sugar in a cell, linked with ordinary burning.

**Source:** https://reasonandscience.catsboard.com/t2139-how-cells-obtain-energy-from-food

Respiration and photosynthesis are two processes that work together. This implies that plant-animal interactions are not entirely one method. Plants, animals, and microbes have coexisted on our planet for far too lengthy several of these have grown inextricably linked to the surroundings of others. Almost all organisms absorb the oxygen created by photosynthesis in the burning of organic molecules (Wendlandt & Stahl, 2015). And many of the $CO_2$ that is bound into organic compounds in a green leaf now was emitted into the air previously by an animal's respiration—or by a fungus or bacteria degrading dead organic matter. As a result, we can see that carbon use is part of a bigger cycle that incorporates the biosphere (any living things on Earth) overall, as well as individual species. Likewise, nitrogen, phosphorus, and sulfur atoms migrate between non-living and living worlds via processes involving plants, animals, fungi, and bacteria (Radi et al., 1991).

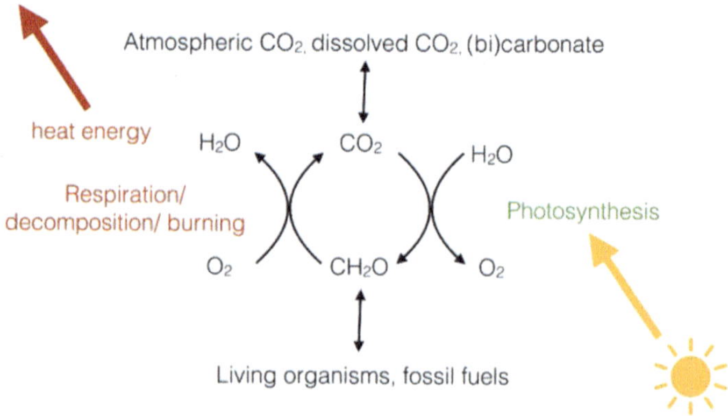


**Figure 6.4.** Graphical Representation of Carbon Cycle.

**Source:** Johnson M. P. (2016). Photosynthesis. Essays in biochemistry, 60(3), 255–273. https://doi.org/10.1042/EBC20160016

### 6.4.1. Oxidation of Molecules

The cell doesn't oxidize organic compounds in a single movement, as occurs in a fire. Metabolism transports particles via a series of processes that seldom require the activities add to oxygen. Before delving into a few of these processes and their motivations, it's necessary to define what the term "oxidation" means (Tsong & SU, 1999).

The term "oxidation," as used here, does not refer only to the adding of oxygens; rather, it refers to any process wherein electrons are shared from one element to the other. In this context, oxidation denotes the elimination of electrons, while reduction denotes the insertion of electrons (Gerischer & Mauerer, 1970). Thus, if $Fe^{2+}$ loses electrons to have become $Fe^{3+}$, it is oxidized, whereas a chlorine atom acquires an electron into becoming $Cl^-$. Because the number of energy in a chemical process is constant (there is no loss or gain), the oxidation process must occur concurrently: that is, if one component acquires an atom in a process (reduction), another molecule emits an electron (oxidation). Whenever a sugar molecule is oxidized to $CO_2$ and $H_2O$, the $O_2$ molecules associated with the formation of $H_2O$ acquire electrons and are called reduced (Parsons & VanderNoot, 1988).

The phrases "oxidation" and "reduction" are applicable sometimes while a partial electron transfer occurs among atoms connected by a covalent bond. Whenever a carbon atom creates a covalent bond with just an atom that has a high electronegativity (Camblor et al., 1992), like oxygen, chlorine, or sulfur, it loses much more than the due proportion of electrons and forms a polar covalent bond: the positive charge on the carbon core is slightly larger than the corresponding charge on its electron density, and the atom obtains a slightly positive charge as well as is seen to be oxidized. In contrast, a carbon atom in a C-H bond possesses significantly more electrons than just its portion and hence is seen to be reduced (Sariaslani & Dalton, 1989).

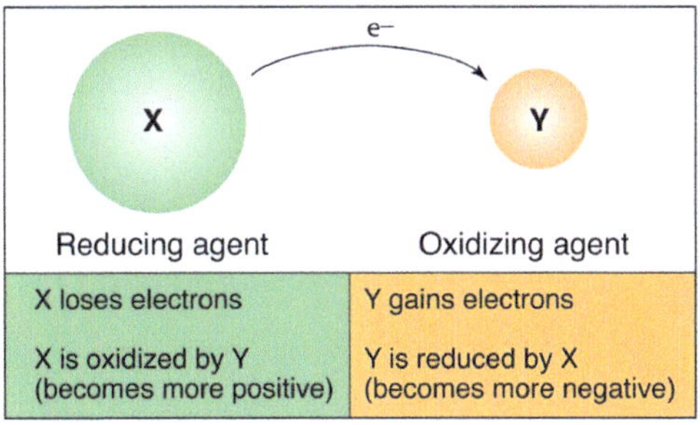


**Figure 6.5.** Reducing and Oxidizing Agents.

**Source:** https://www.kgs.ku.edu/Hydro/GWtutor/Plume_Busters/remediate_refs/redox_chemistry.htm

## 6.4.2. Reduction of Molecules

Whenever two atoms make a polar covalent connection, the atom with the higher proportion of electrons is represented as reduced, whereas the atom with the lower amount of electrons is represented as oxidized.

Whenever a molecule in a cell absorbs an electron ($e^-$), often it absorbs a proton ($H^+$) in the meantime (in water, the proton is presented freely). In this scenario, the overall result is to insert a hydrogen atom into a molecule (Lotze et al., 2007)

$A + e^- + H^+ \rightarrow AH$ (1)

Even though a proton, as well as an electron, are required (rather than simply an electron), these hydrogenation reactions are reduction, while dehydrogenation processes are oxidations. It is very simple to determine if an organic molecule has been oxidized or reduced: reduction occurs when the amount of C-H bonds grows, while oxidation occurs when the quantity of C-H bonds falls. Enzymes are used by cells to catalyze the oxidation process in microscopic stages, allowing useable energy to be obtained. We must now describe how enzymes function and a few of the limits that they face (Cree et al., 2008).

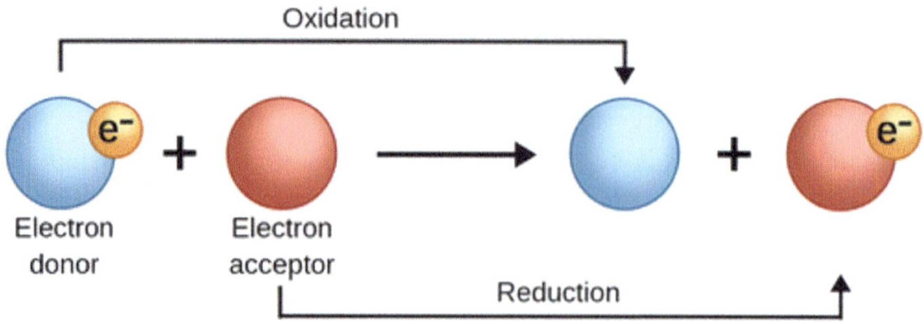


**Figure 6.6.** Redox reactions.

**Source:** https://theory.labster.com/redox-reactions-crl/

## 6.5. ROLE OF ENZYMES IN CHEMICAL REACTIONS

Study the reaction

paper + $O_2$→ smoke+ ashes + heat + $CO_2$ + $H_2O$ (2)

The paper ignites easily and emits energy in the form of heat as well as carbon dioxide and water in the form of gases into the air, but the smoke, as well as dust, don't ever instantaneously recover such objects through the heated environment as well as reassemble themselves into paper. Whenever paper starts burning, its chemical energy is released as heat, which is irreversibly distributed in the arbitrary thermal movements of molecules (Hu & Yang, 2008). In the meantime, the paper's particles and molecules dissipate and then become dysfunctional. There's been a decrease in limitless energy, that's also, the power that can be used to do tasks or start driving chemical reactions, in thermodynamic terms. The decrease of neatness in the manner the power and particles were saved in the paper is reflected in this damage. We'll go over free energy in greater depth later, however, the overall principle is simple: chemical changes only continue in the path that results in the lack of additional energy; like, the sudden orientation for any response is "downhill." In this manner, a "downhill" response is often referred to as "energetically favorable" (Das & Varma, 2010).

A living thing does not vanish in a cloud of smoke, and the book in front of you bursts into flames, even though CO2 is perhaps the most efficient state of carbon under normal circumstances, and $H_2O$ is the most thermodynamically favorable state of hydrogen. This is since the molecules both in the organism as well as the book are pretty stagnant, and they can't be varied to a reduced level without the need for input energy (Benkovic & Hammes-Schiffer, 2003): As a molecule needs a transition state kick across an energy gap beforehand it can undertake a chemical process which wants to leave this in a more steady position. The warm air of a lit match provides the energy barrier in the particular instance of the combustion of a book. An initiator is produced by an abnormally energetic arbitrary collision with neighboring molecules for the particles in the watery solution in a cell—collisions that are becoming more violent as the water is heated (Fong et al., 1976).

The concept of activation energy is very important. Compound X is now in a steady position, and converting it to compound Y requires energy, although Y has a lesser total energy standard than X.

The kick on the energy gap in a living cell is significantly assisted by a specialized class of proteins known as enzymes. So every enzyme attaches tightly to one or two molecules known as substrates as well as retains them in such a manner that the vigor of a chemical process that the bound substances can undertake is greatly reduced (Brameld & Goddard, 1998). A catalyst is a material that lowers the activation energy of a process; catalysts speed up the reactions by allowing a significantly higher percentage of chance collisions with nearby molecules to push the substrates on the energy obstacle. Enzymes are amongst the most powerful catalysts recognized, accelerating reactions by the ratio of 1014 and allowing processes that would not normally take place to take place at average temps (Joshi, 2018).

The chance of a reaction rises dramatically as the energy is reduced. Anywhere at one time, a community of similar substrate particles will have a variety of energies dispersed as illustrated on the graph.

Enzymes are also quite picky. Every enzyme normally catalyzes just one reaction: like, it reduces the activation energy only of that's one of many chemical processes that this is attached substrate molecules may go through. Enzymes do this by directing every one of the cell's numerous distinct chemicals along certain reaction routes (Weiss & Marin, 2008).

Enzyme catalysis analogies in the form of floating balls. (A) To depict enzyme catalysis, a blocking dam is lowered. The capacity of a cell to produce a variety of enzymes, every with its own set of features, is crucial to the survival of living creatures. Every enzyme has a distinct structure that includes an active site, which would be a slot or pocket in the enzyme that only accepts certain substrates. Molecules, like all other catalytic chemicals, stay unaltered after engaging in a process and may therefore be used several times (Raja et al., 2011).

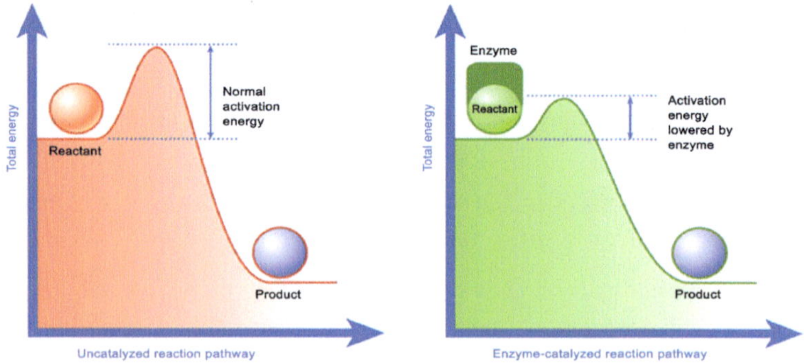


**Figure 6.7.** Enzymes allow activation energies to be lowered.

**Source:** https://www.nature.com/scitable/topicpage/cell-energy-and-cell-functions-14024533/

# REFERENCE

1. Alberts, B., Johnson, A., Lewis, J., Raff, M., Roberts, K., & Walter, P. (2002). Catalysis and the Use of Energy by Cells. In Molecular Biology of the Cell. 4th edition, (pp. 1-5). Garland Science.
2. Armor, J. N. (1999). The multiple roles for catalysis in the production of H2. Applied Catalysis A: General, 176(2), 159-176.
3. Asami, K., Hanai, T., & Koizumi, N. (1980). Dielectric approach to suspensions of ellipsoidal particles covered with a shell in particular reference to biological cells. Japanese Journal of Applied Physics, 19(2), pp. 359.
4. Babcock, G. T., & Wikström, M. (1992). Oxygen activation and the conservation of energy in cell respiration. Nature, 356(6367), 301-309.
5. Bard, A. J. (1979). Photoelectrochemistry and heterogeneous photo-catalysis at semiconductors. Journal of Photochemistry, 10(1), 59-75.
6. Baulig, A., Sourdeval, M., Meyer, M., Marano, F., & Baeza-Squiban, A. (2003). Biological effects of atmospheric particles on human bronchial epithelial cells. Comparison with diesel exhaust particles. Toxicology in vitro, 17(5-6), 567-573.
7. Benkovic, S. J., & Hammes-Schiffer, S. (2003). A perspective on enzyme catalysis. Science, 301(5637), 1196-1202.
8. Beyeler, F., Neild, A., Oberti, S., Bell, D. J., Sun, Y., Dual, J., & Nelson, B. J. (2007). Monolithically fabricated microgripper with integrated force sensor for manipulating microobjects and biological cells aligned in an ultrasonic field. Journal of microelectromechanical systems, 16(1), 7-15.
9. Brameld, K. A., & Goddard, W. A. (1998). The role of enzyme distortion in the single displacement mechanism of family 19 chitinases. Proceedings of the National Academy of Sciences, 95(8), 4276-4281.
10. Brownstein, K. R., & Tarr, C. E. (1979). Importance of classical diffusion in NMR studies of water in biological cells. Physical Review A, 19(6), 2446.
11. Camblor, M. A., Corma, A., Martínez, A., & Pérez-Pariente, J. (1992). Synthesis of a titaniumsilicoaluminate isomorphous to zeolite beta and its application as a catalyst for the selective oxidation of large organic molecules. Journal of the Chemical Society, Chemical Communications, (Vol. 8, pp. 589-590).

12. Centi, G., & Perathoner, S. (2009). Catalysis: role and challenges for a sustainable energy. Topics in Catalysis, 52(8), 948-961.
13. Centi, G., Quadrelli, E. A., & Perathoner, S. (2013). Catalysis for CO 2 conversion: a key technology for rapid introduction of renewable energy in the value chain of chemical industries. Energy & Environmental Science, 6(6), 1711-1731.
14. Cree, L. M., Samuels, D. C., de Sousa Lopes, S. C., Rajasimha, H. K., Wonnapinij, P., Mann, J. R., ... & Chinnery, P. F. (2008). A reduction of mitochondrial DNA molecules during embryogenesis explains the rapid segregation of genotypes. Nature Genetics, 40(2), 249-254.
15. Das, S. K., & Varma, A. (2010). Role of enzymes in maintaining soil health. In Soil enzymology (Vol. 1, pp. 25-42). Springer, Berlin, Heidelberg.
16. Eberhard, S., Finazzi, G., & Wollman, F. A. (2008). The dynamics of photosynthesis. Annual review of genetics, 42(1), 463-515.
17. Evans, J. R. (2013). Improving photosynthesis. Plant Physiology, 162(4), 1780-1793.
18. Farquhar, G. D., Von Caemmerer, S., & Berry, J. A. (2001). Models of photosynthesis. Plant Physiology, 125(1), 42-45.
19. Fong, K. L., McCay, P. B., Poyer, J. L., Misra, H. P., & Keele, B. B. (1976). Evidence for superoxide-dependent reduction of Fe3+ and its role in enzyme-generated hydroxyl radical formation. Chemico-biological interactions, 15(1), 77-89.
20. Gerischer, H., & Mauerer, A. (1970). Untersuchungen zur anodischen oxidation von ammoniak an platin-elektroden. Journal of Electroanalytical Chemistry and Interfacial Electrochemistry, 25(3), 421-433.
21. Gust, D., & Moore, T. A. (1989). Mimicking photosynthesis. Science, 244(4900), 35-41.
22. Hahn, S. I. N. U. H. E., Gehri, R. O. L. A. N. D., & Erb, P. E. T. E. R. (1995). Mechanism and biological significance of CD4-mediated cytotoxicity. Immunological reviews, (Vol. 1, pp. 57-79).
23. Hohmann-Marriott, M. F., & Blankenship, R. E. (2011). Evolution of photosynthesis. Annual review of plant biology, 62(1), 515-548.
24. Hu, H., & Yang, W. (2008). Free energies of chemical reactions in solution and in enzymes with ab initio quantum mechanics/molecular mechanics methods. Annu. Rev. Phys. Chem., 59(1), 573-601.

25. Joshi, R. (2018). Role of enzymes in seed germination. International Journal of Creative Research Thoughts, 6(2), 1481-1485.
26. Ke, B. (2001). Photosynthesis photobiochemistry and photobiophysics (Vol. 10, pp. 2-8). Springer Science & Business Media.
27. Kim, J., Kang, J. W., Park, J. H., Choi, Y., Choi, K. S., Park, K. D., ... & Kim, H. S. (2009). Biological characterization of long-term cultured human mesenchymal stem cells. Archives of pharmacal research, 32(1), 117-126.
28. Lotze, M. T., Zeh, H. J., Rubartelli, A., Sparvero, L. J., Amoscato, A. A., Washburn, N. R., ... & Billiar, T. (2007). The grateful dead: damage-associated molecular pattern molecules and reduction/oxidation regulate immunity. Immunological Reviews, 220(1), 60-81.
29. Loverdo, C., Bénichou, O., Moreau, M., & Voituriez, R. (2008). Enhanced reaction kinetics in biological cells. Nature physics, 4(2), 134-137.
30. Marquez-Curtis, L. A., Janowska-Wieczorek, A., McGann, L. E., & Elliott, J. A. (2015). Mesenchymal stromal cells derived from various tissues: biological, clinical and cryopreservation aspects. Cryobiology, 71(2), 181-197.
31. Parsons, R., & VanderNoot, T. (1988). The oxidation of small organic molecules: A survey of recent fuel cell related research. Journal of electroanalytical chemistry and interfacial electrochemistry, 257(1-2), 9-45.
32. Radi, R., Thomson, L., Rubbo, H., & Prodanov, E. (1991). Cytochrome c-catalyzed oxidation of organic molecules by hydrogen peroxide. Archives of Biochemistry and Biophysics, 288(1), 112-117.
33. Raja, M. M. M., Raja, A., Imran, M. M., Santha, A. M. I., & Devasena, K. (2011). Enzymes application in diagnostic prospects. Biotechnology, 10(1), 51-59.
34. Sariaslani, F. S., & Dalton, H. (1989). Microbial enzymes for oxidation of organic molecules. Critical reviews in biotechnology, 9(3), 171-257.
35. Sariaslani, F. S., & Dalton, H. (1989). Microbial enzymes for oxidation of organic molecules. Critical reviews in biotechnology, 9(3), 171-257.
36. Sato, S., Ishigure, M., & Inaba, H. (1991). Optical trapping and rotational manipulation of microscopic particles and biological

cells using higher-order mode Nd: YAG laser beams. Electronics letters, 27(20), 1831-1832.

37. Tsong, T. Y., & SU, Z. D. (1999). Biological effects of electric shock and heat denaturation and oxidation of molecules, membranes, and cellular functions. Annals of the New York Academy of Sciences, 888(1), 211-232.
38. Wang, F., & Stahl, S. S. (2020). Electrochemical oxidation of organic molecules at lower overpotential: Accessing broader functional group compatibility with electron– proton transfer mediators. Accounts of chemical research, 53(3), 561-574.
39. Weiss, I. M., & Marin, F. (2008). The role of enzymes in biomineralization processes. Met Ions Life Sci-Biomineralization: From Nature to Application. West Sussex, UK: John Wiley & Sons, Vol. 1, 71-126.
40. Wendlandt, A. E., & Stahl, S. S. (2015). Quinone-catalyzed selective oxidation of organic molecules. Angewandte Chemie International Edition, 54(49), 14638-14658.

CHAPTER 7

# STRUCTURE AND FUNCTIONS OF PLANT CELL

CONTENTS

## 7.1. INTRODUCTION

Plant cells are eukaryotic cells present in plants, that are photosynthetic eukaryotes belonging to the Plantae. They feature a membrane-bound nucleus, These include a range of membrane-bound structures that execute a variety of particular activities necessary for the plant cell to operate normally (Showalter, 1993).

Plant cells are the main basis of natural life in all plants species. They are eukaryotic cells, that contain a genuine nucleus and specialized parts called components of cells that complete specific tasks. Chloroplasts are exceptional organelles present in plant cells that are capable of performing photosynthesis to generate carbohydrates. Additionally, they are having cell wall which serves as structural support (Klimecka & Muszyńska, 2007).

Plant cells are often much larger as compared to animal cells, with more comparable dimensions and a cubic or rectangular form. Plant cells also include structural structures not seen in animal cells, such as the cell wall, vacuoles, and plastids, such as the chloroplast. Additionally, animal cells possess features not seen in plant cells, including cilia and flagella, lysosomes, and centrioles (Keegstra, 2010).

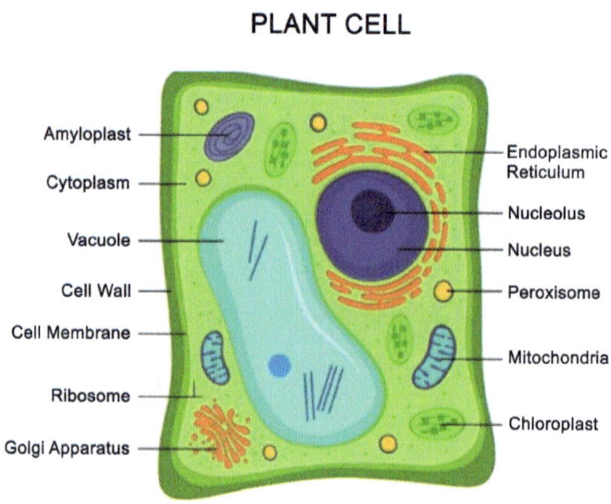


**Figure 7.1.** Structure of a Plant Cell.

**Source:** https://www.shutterstock.com/image-vector/vector-illustration-plant-cell-anatomy-structure-1670413030

The usual properties of a plant cell contain cellulose, hemicellulose, and pectin, plastids that are essential for photosynthesis as well as starch storage, and huge vacuoles which regulate the cell turgidity. Additionally, they have a highly distinct cell division mechanism in which a phragmoplast (composite composed of microtubules, microfilaments, and the endoplasmic reticulum) forms at the time of cytokinesis to divide the daughter cells (Houston et al., 2016).

The majority of such organelles are related to animal cells and serve similar tasks as animal cells. Organelles provide a variety of functions, ranging from the production of enzymes and hormones to the provision of power to a plant cell (Gunning & Steer, 1996).

Plant cells include DNA which aids in the formation of new cells, thereby promoting the plant's development. The DNA is contained inside the nucleus, a membrane-enclosed structure located at the cell's core. Additionally, the plant cell has many organelle complexes that conduct a range of activities necessary for the maintenance of cellular metabolic rates, development, and growth (Ochoa-Villarreal et al., 2012).

Different organelles of a cell will be discussed in the following sections.

## 7.2. PLANT CELL WALL

This is a hard outer shell of a plant that protects the plant cell while also providing its shape (Cosgrove, 2005).

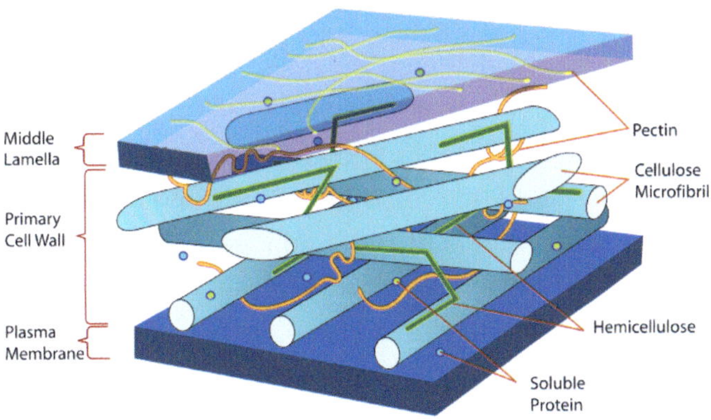


**Figure 7.2.** Diagram of the Plant cell wall.

**Source:** Bailey, Regina. "The Structure and Function of a Cell Wall." ThoughtCo, Sep. 7, 2021, thoughtco.com/cell-wall-373613

It is the stiff outermost layer of the plant that plays a critical function in preserving and shaping the plant cell (Showalter, 1993).

### 7.2.1. Structure of Plant Cell Wall

It's a specific medium that surrounds the plant cell's exterior. Each plant cell contains a cell wall coating, which is a key feature that distinguishes it from animal cells. A central lamella, the main cell wall, as well as a second cell wall are the two sets that make up the cell wall. In between major walls of adjoining cells, the intermediate lamella functions as a reinforcing layer (Cosgrove, 2005).

Underneath the splitting and developing cells is cellulose, which forms the main wall. When contrasted to cells that have attained full maturity, the main wall is much thinner and less stiff. The thinner cell wall permits the cell to grow. Some plants remove the main wall when cell development is complete, while the majority strengthen it or form a secondary wall, which is a stiff layer with a random pattern (York et al., 1986).

The secondary wall provides the plant cell with a constant rigid support structure, similar to that seen in the wood. In comparison to strong secondary walls' intrinsic rigidity and load-bearing capability (Cosgrove, 2016).

### 7.2.2. The Function of the Plant Cell Wall

The cell wall's principal purpose is described as a mechanical and structural job, which is very successful at servicing the plant cell. Such functions involve the following (Showalter, 1993):

- The secondary wall coat provides a protective barrier to the cell while also sheltering it from physically hostile surroundings.
- It is partially permeable, allowing the passage of things like water, molecular nutrients, and minerals into and out.
- It also serves as a stiff key component for the plant, allowing it to construct much of its structures, such as the leaves and stems.
- It also served as a storage location for certain components, like regulatory proteins that identify parasites in the plant, therefore preventing sick tissue from developing.
- Whenever the cell vacuoles are covered with water, the thinner main walls act as structural as well as supporting function levels, generating turgidity on the cell wall, keeping the plants rigid, and avoiding water loss and wilting.

The fundamental building component is composed of cellulose fibers, including main and secondary walls, notwithstanding their composition and structural differences. Cellulose is a polysaccharide layer that provides the cells with strength properties. This power is ingrained in the extremely concentrated water as well as the glycoprotein matrix (Darley et al., 2001).

## 7.3. PLANT CYTOSKELETON

This is a microtubule and filament web which is accountable for preserving the structure of the plant body, as well as offering support for the cytoplasm as well as sustaining its structural structure. Such filaments and tubules ordinarily stretch throughout the cytoplasm of the cell. It also is engaged in the movement of cellular molecules, cell growth, and cell signaling functions, in addition to providing support and sustaining the cells as well as its cytoplasm (Wasteneys & Yang, 2004).

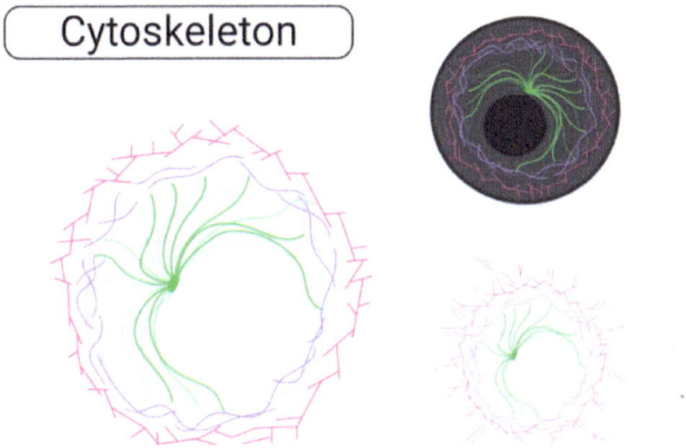


**Figure 7.3.** Structure of cytoskeleton.

Source:https://bio.libretexts.org/Bookshelves/Introductory_and_General_Biology/Book%3A_General_Biology_(OpenStax)/2%3A_The_Cell/04%3A_Cell_Structure/4.5%3A_The_Cytoskeleton

### 7.3.1. Structure of the Plant Cytoskeleton

The cytoskeleton is critical in defining the shape of eukaryotic cells since it describes the cells' support network, maintenance factors, as well as transportation components. These roles are dictated by the cytoskeleton's shape, which is composed of three filaments: microtubules, intermediate filaments, and actin (microfilaments), (Davies et al., 1996). Microfilaments, also called actin filaments, are a network of similar fibers. They are made of fine actin protein strands, thus the term actin filaments. They are indeed cytoskeleton's tiniest filaments, measuring just 7 nanometers in diameter.

- Transitional filaments are around 8-12 nm in size and are located among actin filaments and microtubules. Its role in plant cells is unknown.
- Microtubules are empty tubulin tubing that has a size of 23nm. They are the biggest filaments among the other two (Baluška & Hasenstein, 1997).

## 7.3.2. Functions of the Plant Cytoskeleton

### *7.3.2.1. Microfilaments*

They are primarily responsible for the cytoplasmic partition of the cell, which results in the formation of two daughter cells, a process called cytokinesis. They also take involved in cytoplasmic flowing, which involves cytosol flowing throughout the cell, delivering nutrition and cellular components (Kost et al., 1999).

### *7.3.2.2. Intermediate Filaments*

Although the purpose of middle filaments in plant cells is not fully known, they likely contribute to cell form maintenance, structural strength, and tensile retention inside the cell (Wasteneys & Yang, 2004).

### *7.3.2.3. Microtubules*

Dissimilar to the animal cell, which employs microtubules for cell replication, the plant cell utilizes microtubules to move materials inside the vell as well as to create the plant cell's cell wall (Schmidt & Panstruga, 2007).

In plants, the cytoskeleton also serves the following functions (Shibaoka & Nagai, 1994):

- Providing structure to the plant body, preserving cell structure as well as transporting certain organelle, chemicals, and minerals all through the cell cytoplasm
- It's also involved in cell division during mitosis.
- To summarise, the cytoskeleton is the framework upon which the cell is built, and it so sustains, supports, and characterizes the cell shape.

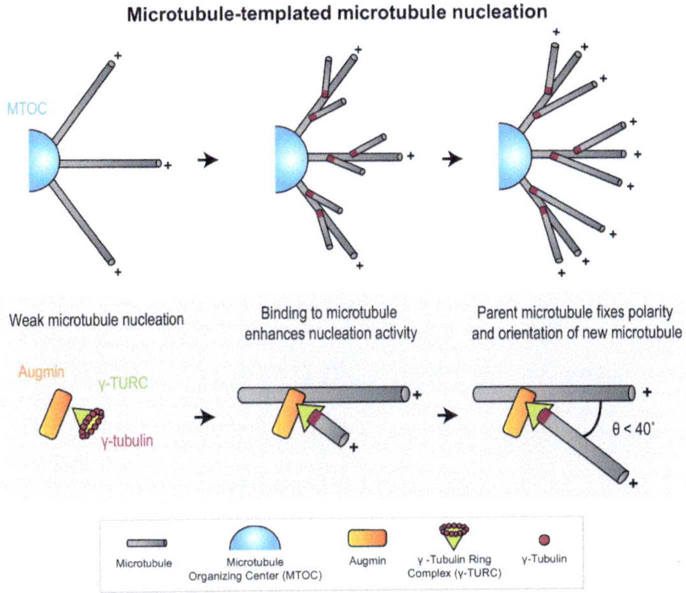


**Figure 7.4.** Structure of microtubules.

**Source:** eLife 2021;10:e63880 *DOI:* 10.7554/eLife.63880

# 7.4. PLANT CELL MEMBRANE

The cell membrane is found in plants among the cytoplasm as well as the cell wall. As with animal cells, the cellular membranes of plants have two layers of lipids. It is composed mostly of phospholipids but also includes proteins and carbohydrates (Gonzalez & Barrett, 2010).

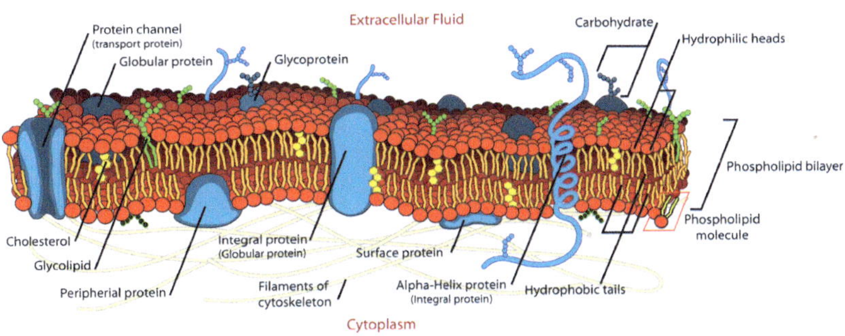


Figure 7.5. Plant cell membrane.

**Source:** https://www.microscopemaster.com/cell-membrane.html

### 7.4.1. Structure of the plant cell (plasma) membrane

The plasma membrane, similar to other cell membranes, is composed of proteins and lipids. A phospholipid bilayer, which creates a tough boundary among two liquid cores, is the membrane's main arrangement. Such divisions are inner plus outer of a cell within an instance of the plasma membrane. The plasma membrane's specific activities, like specialized molecular transit and cell-cell identification, are performed by structural proteins within the phospholipid bilayer (Schindler et al., 1989).

### 7.4.2. Functions of the Plant Cell (Plasma) Membrane

The activities of the plant cell are as follows (Hamann, 2015):

- In-plant cells had the cytoplasm and the cell wall divided by the cell membrane.
- Because it has a specific permeability, it can govern the materials that enter and exit the cell.
- Additionally, it shields the cell against exterior injury and gives it balance and comfort.
- It contains embedded proteins that are coupled along with lipids and carbohydrates and utilized to move biological components over the membrane.

## 7.5. PLASMODESMATA OF THE PLANT CELL

These are tiny passageways that aid in communication as well as material movement between plant cells. They link the plant's cell compartments, facilitating internal cell transport of nutrients, water, minerals, as well as other substances. They also enable biological molecules to communicate with one another. Plasmodesmata are classified into two categories (Sevilem et al., 2013)

- **Primordial plasmodesmata**, develop after cellular division.
- **Secondary plasmodesmata** occur within matured plant cells.

Whenever the latest cell wall is produced throughout cellular division, the portion of the endoplasmic reticulum gets trapped in the central lamella and forms the main plasmodesmata. They develop a link with one another as they form, and at the connecting location, they produce thin gaps called holes on the sides. Secondary plasmodesmata are plasmodesmata that have been introduced into adult cells only among their cell walls. These are present in the plant as well as algal cells and have evolved separately. The shape of plasmodesmata is governed by a callose polymer generated throughout cell cytokinesis (Knox & Benitez-Alfonso, 2014).

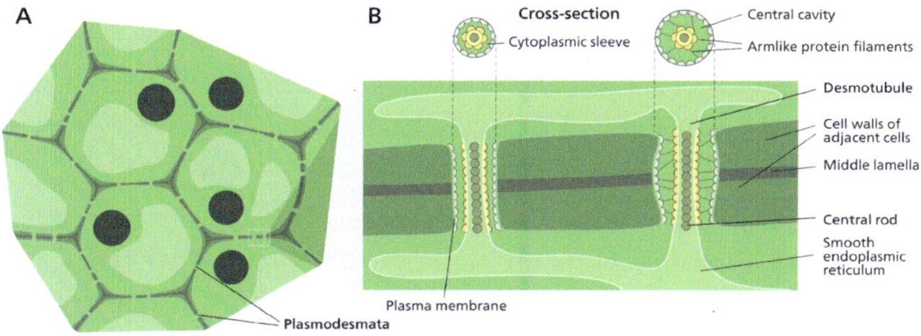


**Figure 7.6.** Diagram of Plasmodesmata.

**Source:** https://www.news-medical.net/life-sciences/What-are-Plasmodesmata.aspx

## 7.5.1. Structure of Plasmodesmata of Plant Cells

Plasmodesmata measure between 50 and 60 nm in size. They are composed of three layers, namely the plasma membrane, the cytoplasmic cuff, and desmotubules. These coatings may increase the thickness of the cell wall by the range of 90nm (Zambryski, 2004).

- **The plasma membrane**-- is the constant expansion of the plasmalemma composed of several layers of phospholipids.
- **Cytoplasmic sleeves** — are liquid pockets that are contained by the plasmalemma, producing an infinite loop of cytosol.
- **Desmotubules** –this is a flat tube that originates in the endoplasmic reticulum and connects 02 neighboring cells.

## 7.5.2. Functions of the Plasmodesmata

Transcript proteins, small pieces of RNA, mRNA, viral genomes, as well as virus-related molecules are all transported from one cell to the other. For example, the MP-30 proteins of the tobacco mosaic virus attach to the viral genome and transport this from affected to unaffected cells via the plasmodesmata. MP-30 is hypothesized to attach to the virus's genetic genome and transport it via plasmodesmata between affected to unaffected cells. With the support of partner cells, they are employed to control sieve tube cells. They're also employed by phloem cells to make nutrition delivery easier (Cilia & Jackson, 2004).

## 7.6. THE CYTOPLASM OF THE PLANT CELL

It is a kind of gel substance that lies directly underneath the cell membrane and contains the majority of the cell's contents. Water, enzymes, salts, organelles, plus different organic compounds comprise it.

It isn't classed as among the cell's organelles since it just serves as a physical link to retain and store most of the complicated cell's internal organelles, as well as conveying and regulating cell chemicals to keep the cell alive (Cairns et al., 1978).

It is since few of such organelles get their protective membranes; for instance, the mitochondria and Golgi bodies have had at minimum two layers that serve a variety of purposes (Streward, 1942).

The nuclei are not included in the cytoplasm due to its double-level core structure and the presence of its organs and sub-organ.

Central vacuoles, lysosomes, endoplasmic reticulum, storage granules, plastids, Golgi bodies, and mitochondria, are all found in the plant cytoplasm (Brewin, 2004).

## 7.7. PLASTIDS OF PLANT CELLS

Plastids are unique organelles present just in the cells of plants and algae plants. They are characterized by the double membrane. They feature distinctive colors that enhance their systems, particularly in the storage and processing of food. Such chemicals also contribute to the plant's color (Bock, 2001).

Plastids are a double-membrane organelle present in the body of algae and plants that are utilized to generate and retain food. Plastids get the capacity to discriminate between their types and may reproduce quickly by binary fission, creating around 1000 plastid duplicates depending on the cell. Plastids are reduced to roughly 100 per matured cell in mature cells (Lopez-Juez & Pyke, 2004).

Plastids are derived from proplastids (indistinct plastids), which are present in the plant's meristematic tissues.

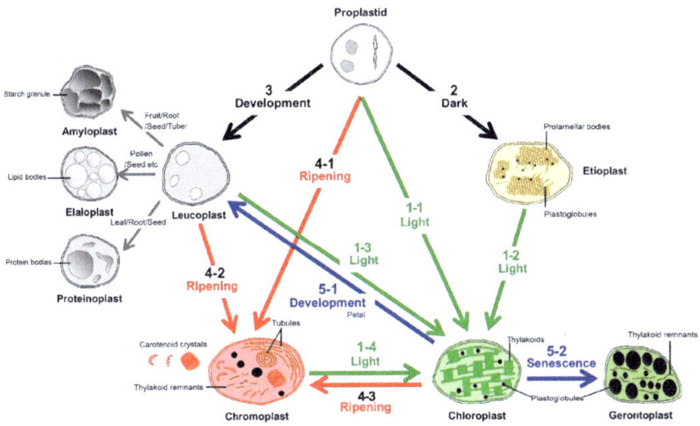


**Figure 7.7.** Diagram of types of plastids.

**Source:** Choi H, Yi T and Ha S-H (2021) Diversity of Plastid Types and Their Interconversions. Front. Plant Sci. *12:692024.* doi: 10.3389/fpls.2021.692024

## 7.7.1. Development of Plastids

Plastids are enormous protein-DNA complexes that are connected with the cell's intermembrane space and are termed plastid nucleoids. There are at minimum 10 duplicates of plastid DNA in the nucleoids. Proplastids are plastids that are homogeneous and have just one nucleoid. These develop into plastids, that have additional nucleoids at the membrane's borders that are attached to the internal envelope membrane (Thomson & Whatley, 1980).

The proplastid nucleoid experiences modification throughout development and differentiation, altering its form, and length, and moving to a new place inside the organelle. The nucleoid proteins are involved in this restructuring pathway (Caredda et al., 1999).

The following are among the roles of plastids (Longo et al., 1979):

- Since chlorophyll pigment is present in the chloroplast, so they are directly engaged in the production of food for the plant via photosynthesis.
- They also hold starch as a source of nourishment.
- They get the capacity to provide power for the cell's systems by synthesizing fatty acids and terpenes.
- Palmitic acid, a chloroplast-produced ingredient, is utilized to make the plant cuticle and waxy substances.

### 7.7.2. Types of Plastids

The classification of plastids is dependent on their activities as well as the existence of distinctive colors. They involve the following (Schnepf, 1980):

- **Chloroplasts** — photosynthesis-related green plastids
- **Chromoplasts** - pigmented plastids that are employed by plants to synthesize and retain pigments
- **Gerontoplasts** — degrade the photosynthetic machinery of plants as they age.
- **Leucoplasts - are** translucent plastids that are employed to synthesize terpene compounds that give protection. They may differentiate, resulting in the formation of specialized plastids with a variety of roles (Voigt et al., 2010).

## 7.8. PLANT CELL CHLOROPLAST

Chloroplasts are plant organelles that are using photosynthesis energy to transform energy from the sun into pretty steady chemical energy. By doing this, they keep life on Earth going. Chloroplasts also play several metabolic functions for the plant cells, like the formation of fatty acids as well as lipid membranes (McFadden, 2001).

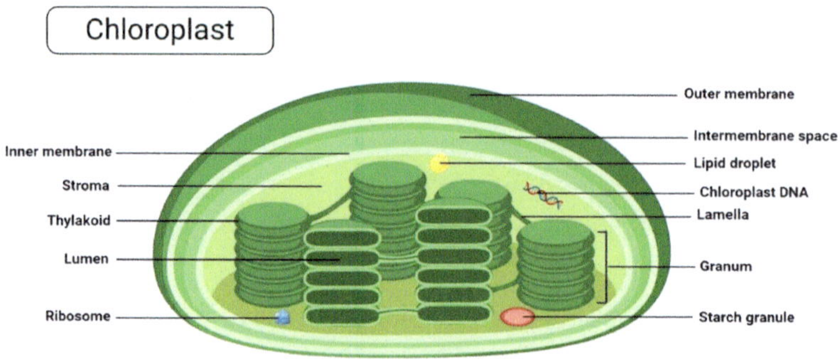


**Figure 7.8.** Diagram of chloroplast.

**Source:** https://www.britannica.com/science/chloroplast

### 7.8.1. Structure of the Plant Cell Chloroplast

The chloroplast arrangement of a plant cell is described here (Goldschmidt-Clermont, 1997):

- These organelles may be present in both plants as well as algal cells.
- They have an oval form.
- They are comprised of two external membranes (external and internal), as well as two membranes in an interior structure called the thylakoid membrane.
- The top layer creates the chloroplast's exterior coating, whereas the internal membrane lies underneath it.
- The membranes are surrounded by a thin membrane gap, as well as the stromal is a compartment inside the membranes. The chloroplast is housed in the stroma.
- The thylakoid membrane is widely stretched, leading to the presence of a flat ring called thylakoids, who contain vast amounts of chlorophyll and carotenoids, as well as the active transport, also called the light-harvesting complexes, that is employed throughout photosynthesis.
- Thylakoids are layered on top of one another to form grana (Weier, 1938).

### 7.8.2. Functions of the Plant Cell Chloroplast

The following are the tasks of plant cell chloroplast (Levy et al., 2015):

- The chloroplast is the location of food manufacture for plant cells, through a method process called photosynthesis.
- Chloroplasts have chlorophyll, a green-colored chemical that collects solar light sunlight for photosynthesis.
- Photosynthesis turns the water, carbon dioxide, plus radiant energy into nutrients that plants can use.
- Thylakoids include chlorophyll pigment as well as carotenoids, which help to capture energy from the sun to be used in the photosynthetic process.
- The chlorophyll pigment is responsible for the green hue of plants.

## 7.9. CHROMOPLAST PLASTID OF THE PLANT CELL

Chromoplasts are responsible for storing and synthesizing all of the pigments found in plants. They occur in a wide range of plants of varying years. They are often created from chloroplasts, which is the label applied to a place in the plant where all the colors are stored and manufactured. They contain carotenoid colors which contribute to the color distinction found in flowers and fruits. Its hue invites bees and their pollination systems (Sadali et al., 2019).

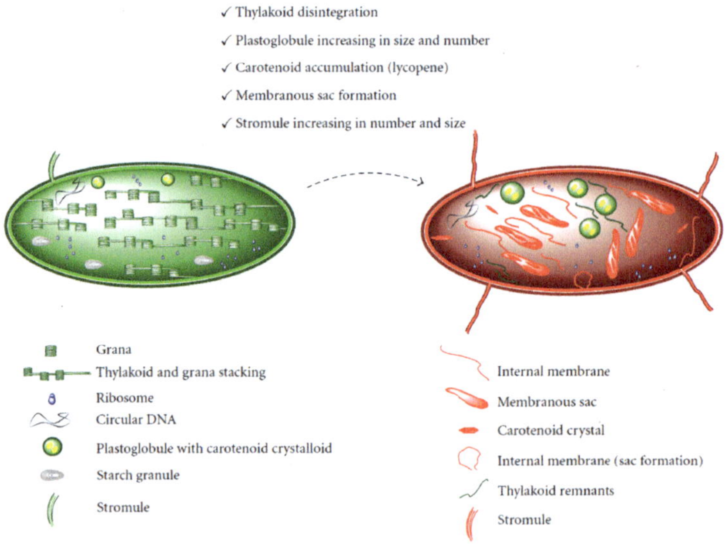


**Figure 7.9.** Diagram of chromoplast.

**Source:** Wanping Bian, Cristina Barsan, Isabel Egea, Eduardo Purgatto, Christian Chervin, Mohamed Zouine, Alain Latché, Mondher Bouzayen, Jean-Claude Pech, "Metabolic and Molecular Events Occurring during Chromoplast Biogenesis", Journal of Botany, vol. 2011, Article ID 289859, 13 pages, 2011. https://doi.org/10.1155/2011/289859

## 7.9.1. Structure of Plant Chromoplast

According to microscopic examination, there are at minimum four varieties of chromoplast (Camara et al., 1995):

- Granule-containing proteic stroma.
- Granular amorphous pigment.
- Crystals of protein and pigment
- Chromoplasts that have crystallized.

However, a more particular characteristic has been identified, and it has been classified into five kinds:

- Globular chromoplasts are globular chromoplasts that look like globules (Pozueta-Romero et al., 1997).
- A chromoplast that looks to be crystallized.
- Fibrillar chromoplast (fiber-like chromoplast).
- Tubular chromoplast (tube-like chromoplast).
- Membranous chromoplas.

A few plants have particular forms of chromoplasts, like mangoes, which contain globular chromoplasts, whilst also carrots have solidified chromoplasts, and tomatoes contain both crystalline as well as membranous chromoplasts since they pile up carotenoids.

### 7.9.2. Functions of Plant Chromoplast

The activities of plastids in plant cells are described here (Tan et al., 2019).

- They provide plant components including flowers, fruits, roots, and leaves in different hues. The transition from chloroplast to chromoplast causes the plant's fruits to mature.
- They make and retain plant pigments like yellow colors for xanthophylls and orange colors for carotenes. The color of the plants and their sections comes from this.
- The pigments they generate invite pollinators, that aid in the replication of the plant pollen.
- Chromoplats in roots allow water-insoluble substances to accumulate, notably in tubers like potatoes and carrots.
- They have a role in floral, fruit, and leaf pigmentation that alters as plants age.

### 7.9.3. Gerontoplast Plastids of the Plant Cell

Such plastids, which are present in plant leaves, are the components of the cell that cause cells to age. They separate from chloroplasts while plants mature and lose their ability to produce photosynthesis. They show as unstacked chloroplasts devoid of a thylakoid membrane as well as a buildup of plastoglobuli responsible for the cell's energy production (Zavaleta-Mancera et al., 1999).

Gerontoplast's principal role is to help in the degeneration of plant components by providing them a unique hue to signify a loss of photosynthesis.

### 7.9.4. Leucoplast Plastids of the Plant Cell

Such are the plastids that aren't colored. They are present in non-photosynthetic portions of plants, such as the stems as well as seeds, because they have no chloroplast colors (Turner et al., 1999).

They are shorter than chloroplasts, which have a variety of morphologies, some of which resemble ameboid shapes. They are linked by a web of stromules in the roots and flower petals.

They may be adapted to retain vast amounts of starch, lipids, or proteins, and are therefore called amyloplasts, elaioplasts, and proteinoplasts, based on what they retain (Brunkard et al., 2015).

The leucoplast's primary role is to:

- Starch, lipids, and protein preservation
- Amino acids, as well as fatty acids, also are converted using them.

## 7.10. PLANT VACUOLES

Plant vacuoles are huge chambers that account for a substantial portion (up to 90%) of the size of plant cells. Water will move into the vacuole under normal development circumstances, increasing the turgor pressure that causes cell wall enlargement (Marty, 1999).

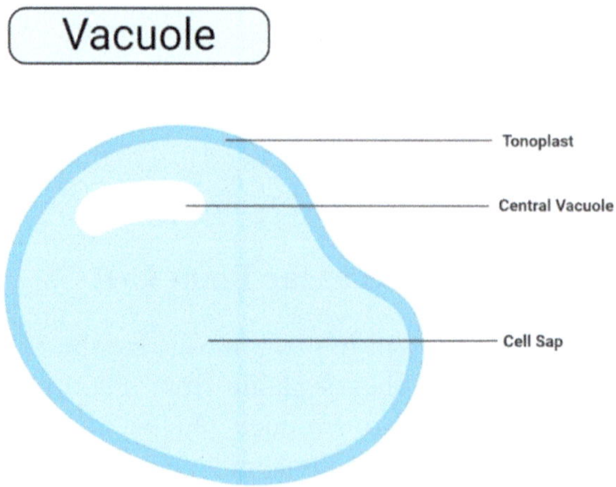


**Figure 7.10.** Structure of Plant vacuoles.

**Source:** https://www.nature.com/scitable/topicpage/plant-vacuoles-and-the-regulation-of-stomatal-14163334/

The following diagram illustrates the composition of plant vacuoles:

- In comparison to animal cells, plant cells feature huge vacuoles.
- Central vacuoles are present in the cytoplasmic membrane of cells of a wide range of animals but are bigger in plant cells.
- These are huge, fluid-filled vesicles found inside a cell's cytoplasm.
- It comprises 30% of the cell's fluid size but may occupy close to 90% of the cell's intracellular compartment.

## 7.11. MITOCHONDRIA OF THE PLANT CELL

Plant mitochondria are complicated organelles that play several metabolic activities such as the generation of energy for cell activation, and the production as well as the destruction of several chemicals (Lam et al., 2001).

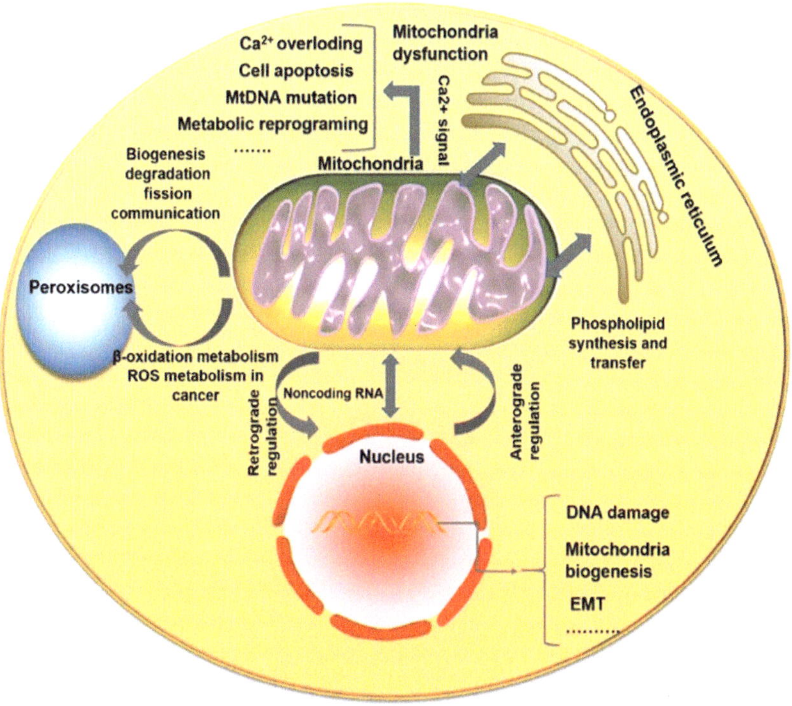


**Figure 7.11.** Structure of Mitochondria.

**Source:** Xia, M., Zhang, Y., Jin, K. et al. *Communication between mitochondria and other organelles: a brand-new perspective on mitochondria in cancer.* Cell Biosci **9,** 27 (2019). https://doi.org/10.1186/s13578-019-0289-8

### 7.11.1. Plant Cell Mitochondria Definition

Mitochondria called chondriosomes are the energy-generation cell organelles, consequently, they are generally recognized as the energy for the cell. The mitochondria transform deposited nutrients with the assistance of oxygen to create power in for of (ATP ) Adenosine TriPhosphate, therefore they are the location for non-photosynthetic power transmission (Mackenzie & McIntosh, 1999).

There are dozens of mitochondria inside one plant cell. Mitochondria are seen in great quantities inside the phloem color of the plant cell, and the surrounding cells have great metabolic levels. This is to offer energy that sustains numerous needed processes, such as the movement of food via the sieve elements. As they conduct functions, mitochondria continually travel and alter their forms, based on their contacts with light captured for photosynthesis, the number of cytosolic carbohydrates as well as even the endoplasmic reticulum induced interaction (Hammani & Giegé, 2014).

The plant and animal mitochondria are the same excluding some noticeable changes e.g. mitochondria in plants have decreased nicotinamide adenine dinucleotide (NADH) dehydrogenase employed for oxidation of exogenous NADH that animal cells lack.

Mitochondria in numerous plant bases are generally tolerant to cyanide suppression, a trait not observed in animal mitochondria. And from the opposite side, the b -oxidation route of fatty acids is situated in animal mitochondria, but in plants, the catalysts of fatty acid oxidation occur in the glyoxysomes (Robson & Vanlerberghe, 2002).

### 7.11.2. Structure of Plant Mitochondria

The following points depict the composition of plant mitochondria (Andre et al., 1992):

- The pleomorphism of plant cell mitochondria is significant.
- Mitochondria are distinct, spherical-oval-shaped structures with diameters varied from 0.2 to 1.5 m in vegetation.
- Mitochondria have a two-layered structure, with a clean outer surface and a complicated inner layer enclosing the organelle core.
- Lipid bilayers containing a hydrophobic fatty acid sequence form the two levels. These phospholipids belong to a group

of phospholipids that are extremely dynamic and have a great affinity for fatty acid regions.

- In the core bulk, there is a mitochondrial gel matrix.
- The Tricarboxylic Cycle (TCA) enzymes citrate synthetase, pyruvateoxidase,isocitratedehydrogenase,malatedehydrogenase, and malic enzymatic are all found in mitochondria.

### 7.11.3. Functions of Mitochondria in Plants

The following sections discuss the actions of plant mitochondria (Mackenzie & McIntosh, 1999):

- Because the mitochondria are the cell's powerhouse, their primary role is to generate energy for utilization by the cell.
- To get a great metabolism rate since they provide energy for an unknown process that transports meals, mostly sucrose, via sieve tubes.
- In mitochondria, the energy stored in photosynthesis-produced food is utilized to power the cells' metabolic rates. The mitochondria, for instance, provide energy for the synthesis of new cells, enzyme formation, and the movement of sugar molecules.
- This is the citation for the tricarboxylic acid cycle (TCA), sometimes referred to as the Krebs cycle. The TCA cycle utilizes the nutrition in the cell, transforming them into wastes however that mitochondria may utilize to generate energy. These events occur in the intermembrane space as a result of the membrane's bending into folding termed cristae, which contain the protein parts of the primary energy generation system cells, referred to as the Electron Transport Chain (ETC). ETC is the primary generator of ATP in the body (Costes & Vartapetian, 1978).

## 7.12. ENDOPLASMIC RETICULUM (ER) OF THE PLANT CELL

The endoplasmic reticulum (ER) is a constant web of folded membrane sacs located inside the cytoplasm of the cell. It is a complicated organelle that consumes a substantial portion of the cytoplasm of the cell. It is divided into two areas: the coarse endoplasmic reticulum (which contains ribosomes) and the smoother endoplasmic reticulum (which lacks ribosomes) (they miss ribosomal linkage) (Deng et al., 2013).

The endoplasmic reticulum, which is well-known for its high-speed operations in eukaryotic cells, is involved in the synthesis, processing, transport, and storage of proteins, lipids, and chemical components. The plant cell as well as other structures like the vacuoles and apoplast utilize these components (Plasma membrane) (Wan & Jiang, 2016).

- The lumen is the term used to describe the inside region of the ER. The inner space of the ER is known as the lumen.
- It is linked to the nuclear envelope, connecting the nucleus to the cytoplasm of the cell and also connecting the cell to the plasmodesmata connecting to the plant cells. It comprises 10% of the cytosol's volume.
- Rough ER, contrary to this, usually invariably occurs as stacking of double membranes densely packed with ribosomes. Rough ER, depending on its constant look, most probably comprises parallel membrane sheets instead of the tubular layers seen in smooth ER.
- Cisternae, or cisternal cells, are flat, linked vesicles. The rough ER's cisternal cells are indeed known as luminal cells. Therough endoplasmic reticulum and the Golgi complex are made of cisternal cells (Matsunaga et al., 2013).

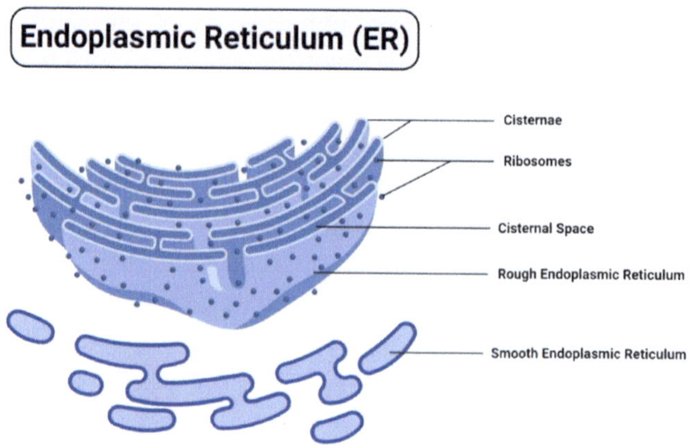


**Figure 7.12.** Structure of Endoplasmic reticulum (ER).

**Source:** https://www.britannica.com/science/endoplasmic-reticulum

### 7.12.1. Structure of plant cell endoplasmic reticulum

The following defines the design of a plant cell (Brandizzi, 2021):

- This is a regularly foldable membranous organelle located in the cell cytoplasm that is composed of a small web of flattening linked divisions (sacs) that links the cytoplasm to the nucleus of the cell.
- Its membranes include membranous gaps known as cristae spaces, and the membrane folds are referred to as cristae.
- Rough Endoplasmic Endoplasmic Reticulum (ER) and Smooth Endoplasmic Reticulum (SER) are the two main forms of ER depending on their performance and structure.

### 7.12.2. Functions of the Endoplasmic Reticulum

Ribosomes surround the endoplasmic reticulum's outer membrane, giving it a tough lumpy look. The Rough ER's principal function is to synthesize proteins that are then taken from cells to the Golgi bodies, that transfer them to various sections of the plant to aid in development. Such proteins are formed by the combination of amino acid composition to generate antibodies, hormones, and digestive enzymes. The ribosomes connected to the crude ER assemble (Di Sano et al., 2012).

While a few proteins are synthesized on the outer side of the cell, they may also be carried into the Rough ER, in which they are assembled into the appropriate form and structure for cell use and coupled with sugar components to create a full protein. Such combinations are subsequently carried and disseminated to regions of the ER called the transitional ER, where they are packaged in cell vesicles and exported to other areas of the plant through the Golgi bodies (Fenech et al., 2020).

The smooth Endoplasmic reticulum is smooth owing to the absence of external ribosomes connected. They seem to be emerging from the rough lumen. Its function is to synthesize, secrete, and store lipids, as well as to metabolize carbohydrates and manufacture new membranes. This is defined by the existence of multiple enzymes on its exterior (Germain et al., 2002).

Whenever a plant has sufficient energy for photosynthesis yet has surplus lipids produced by the cells, these lipids are retained as triglycerides in the endoplasmic reticulum ( er. And then when the cell requires additional energy, the triglycerides are decomposed to provide the plants with the energy they need. At the most basic level, an endoplasmic reticulum has been associated with the synthesis of cellulose on the cell wall (Petrova et al., 2020).

### 7.12.3. Other Functions of the Endoplasmic Reticulum in the Plant Cell

Calcium is needed for the development and growth of plant cells, which promotes plant growth. However, in rare situations, calcium is created in surplus, leading the plant cell to die. As a result, the Endoplasmic reticulum has indeed been implicated in calcium regulation by turning it into calcium oxalate crystals. Crystal idioblasts, cells in the endoplasmic reticulum, perform a critical part in this transformation and in preserving such crystals (Di Sano et al., 2012).

Additionally, the ER serves as a plant sensor. Plants may respond rapidly to environmental stimuli such as intensity of light, temperature, and air pressure. The ER acts as a mediator in these systems, allowing the plant to react appropriately. For instance, the Venus flytrap plant is very delicate. It is because it has cortical endoplasmic reticulum (Cortex cells) that react rapidly to make contact.

Sensation causes the sensory ER to migrate and congregate at the bottom and top of the cell, squeezing themselves together and imposing a limit on themselves. This results in the emission of calcium that has accumulated, which provides the sensation of touch (Chen et al., 2012).

The plasmodesmata and the cortical ER are inextricably intertwined (a narrow thread of cytoplasm that moves across the cell walls of neighboring plant cells and permits connection between them). The Plasmodesmata serves as a communication route between cells, connecting to the motor cells and causing the cells and plant to react appropriately (Ozcan & Tabas, 2012).

## 7.13. RIBOSOME OF PLANT CELL

It is the organ system in charge of the cell's protein production. It is abundant in the cell cytoplasm, and several of them, known as functional ribosomes, may also be located in the nuclei, mitochondria, and the chloroplast. Ribosomal DNA (rDNA) and cell proteins make up its structure. The translation is the procedure by which ribosomes synthesize proteins by employing messenger RNA, that transports nucleotides to the ribosomes. The message is then guided and translated by the ribosomes in the type of nucleotides provided by the mRNA (Byrne, 2009).

**Figure 7.13.** Structure of Ribosome.

**Source:** https://www.microscopemaster.com/ribosomes.html

## 7.13.1. Structure of Ribosomes of the Plant Cell

Ribosomes have a similar structure in all cells or are shorter in bacterial cells. Ribosomes in eukaryotic cells are typically big and should only be quantified in Svedberg units (S). On centrifugation, the S unit is used to quantify the accumulation of big molecules into particles. A large S value indicates a rapid deposition rate, implying a higher mass (Scharf & Nover 1987).

In the 1990s, eukaryotic cell sediment predominated, whereas prokaryotic cell sediment predominated in the 1970s. The mitochondrial and chloroplast ribosomes are identical in size to the bacterial ribosomes. Generally, ribosomes are composed of two components, namely small and big subunits, which are classed by the S unit as per their rates of sedimentation. As a eukaryotic cell, the plant cell possesses enormous complicated ribosomes with increased S units, four RNAs, and over 80 proteins. The big subunit contains 42 proteins from the S subunit of the sixties (28S RNA, 5.8S RNA, and 5S RNA). The tiny subunit sediments at the ratio of 40S are composed of 01 RNA and 33 proteins. The ribosome assembles in the cell's nucleolus and is subsequently transferred into the cytoplasm via nuclear holes. Protein synthesis takes place mostly in the cytoplasm (translation) (Weis et al., 2015).

### 7.13.2. Functions of ribosomes in plant cells

The following are the roles of plant ribosomes (Warner et al., 2009):

- Ribosomes, which include an RNA component, are responsible for synthesizing proteins for biological tasks like cell repair.
- Ribosomes act as catalysts in producing strong binding for portion extension using peptidyl transfer and peptidyl hydrolysis.
- Ribosomes, which are present in the cytoplasm of cells, are capable of converting genetic instructions into a sequence of amino acids as well as assembling protein polymers from monomers of amino acids.
- They're also employed in protein folding and assembly.

## 7.14. STORAGE GRANULES OF PLANT CELL

Such kind of clumps is seen in the cytoplasm and plastids of plant cells. They are inactive cells present in plants that serve primarily as a storage compartment for starch. They serve as food storage containers. They retain carbs in a kind of glycogen or carbohydrate polymer for the cell.

They retain starch granules for the plant body organically. Additionally, they power metabolic processes inside the cell that include chemical reactions, providing energy for the synthesis of new cellular components (Kiss et al., 1988).

## 7.15. GOLGI BODIES OF PLANT CELL

The Golgi complex also referred to as the Golgi apparatus, is a complicated membrane-bound part of the cell present in the cytoplasm of eukaryotic cells. They are located around the nucleus, right adjacent to the endoplasmic reticulum (DaSilva et al., 2004).

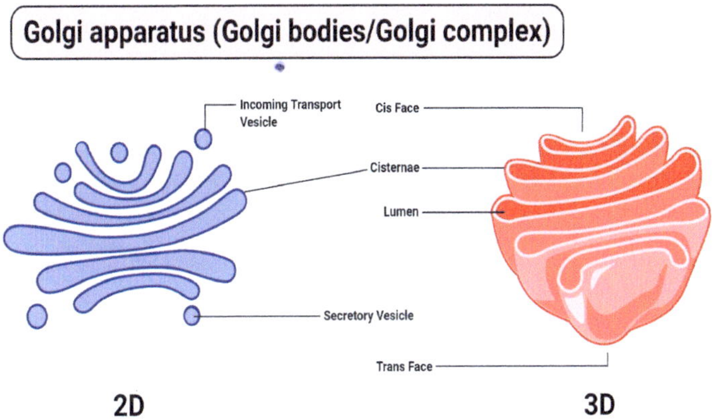


**Figure 7.14.** Structure of Golgi apparatus.

**Source:** https://www.britannica.com/science/Golgi-apparatus

## 7.15.1. Structure of the Golgi Bodies in a Plant Cell

The following defines the design of the Golgi bodies (Hawes et al., 2008):

- Golgi bodies are held collectively by cytoplasmic microtubules also gripped by a protein matrix.
- They are composed of flat stacking sacks called cisternae.
- Plant cells contain thousands of Golgi bodies traveling all along the cytoskeleton, across the endoplasmic reticulum, as opposed to just a handful in animal cells (1-2).
- The Golgi bodies are divided into three major portions (Akkerman et al., 2011):
  - **The Cis Golgi system** often called the Goods inwards cisternae, are the cisternae nearest towards the endoplasmic reticulum. This is the entrance point to the Golgi body and is also known as the cis Golgi reticulum.
  - **The middle of the Golgi stack-** this is the primary processing center, located in the cisternae's core layer.
  - **The Trans Golgi network is** sometimes referred to as the Goods out cisternae. That's the cisternae endoplasmic reticulum that is the furthest away from the endoplasmic reticulum.

### 7.15.2. Functions of the Golgi Bodies in a Plant Cell

The Golgi bodies provide a variety of tasks, including being a neighboring organelle to the endoplasmic reticulum and delivering cell outputs. They are present as a membrane structure in the center of the cell's pathway, largely processing, distributing, and storing proteins for usage by the plant at the time of stress reactions and many others in leguminous plants like grain foods.

The membranous sac chambers provide a variety of chemically linked tasks. As new proteins move through the Golgi bodies on their way out from the endoplasmic reticulum, they move through the three positions, each of which performs a separate reaction on the molecule, changing them in various ways, such as (Northcote & Pickett-Heaps, 1966):

- Chunking the protein molecules to oligosaccharides links
- Sugar moieties from various side chains are attached to protein components.
- Fatty acids, as well as phosphate groups, are added to the components, while monosaccharides are removed.

The cell vesicles convey protein molecules from the endoplasmic reticulum into the cis sector, where they are changed and packed before being transported to the next chamber. The vesicle's transit is aided by the addition of a tag, such as a phosphate set or specific protein molecules, which directs it to the next destination (Mollenhauer et al., 1961).

After the proteins, as well as lipid molecules, have been delivered by the vesicles, the Golgi bodies are in charge of synthesizing the output and conveying it to its ultimate destination. The existence of enzymes in the Golgi bodies of vegetation aids this process by attaching sugar moieties to proteins, packaging them, and delivering people to the outer membrane (Martinez et al., 1994).

## 7.16. THE NUCLEUS OF THE PLANT CELL

The nucleus is the cell's information processing core. It is a very sophisticated special organelle whose major duty is to preserve the genetic data of the cell. Additionally, it is capable of managing the cell's functions, which include metabolism, growth, protein and lipid synthesis, and, more broadly, cell reproduction through cell division processes (Deslandes & Rivas, 2011). On the Chromatids, the nucleus carries the cells' genetic data, referred to as Deoxyribonucleic Acid (DNA) (specialized kind of threads strands of nucleic acids plus protein present in the nucleus, having genetic data) (Lacroix et al., 2006).

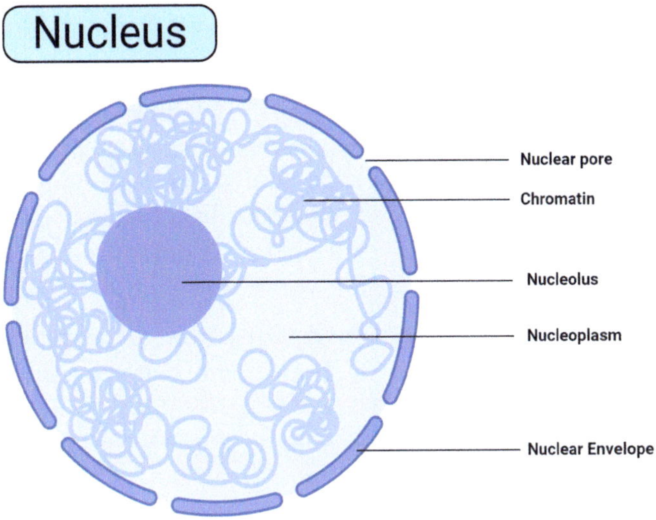


**Figure 7.15.** The nucleus of the Plant Cell.

**Source:** https://biologydictionary.net/cell-nucleus/

## 7.16.1. Structure of the Nucleus of the Plant Cell

The nuclei of a plant cell have the following format (Sakai et al., 2004):

- It is cylindrical structured and located in the cell's center, taking up 10% of cell size.
- It has a double-layered wall called the nuclear envelope that divides the contained in the nucleus from that of the cytoplasm of the cell.
- Chromatins, DNA (that produces cell chromosomes at the time of cell division), and the nucleolus (that also synthesizes cell ribosomes) were among the nuclear components.

## 7.16.2. Functions of the Nucleus of the Plant Cell

The primary purpose of the cell nucleus is to serve as the command center for the cell. The nuclear membrane is present; it separates the nucleus along with its components from cytoplasmic organelles. This nuclear membrane is surrounded by the nuclear cover, which has many nuclear holes and allows for permeation between the cytoplasm and the nucleus (Kaiserli & Jenkins, 2007). A web of microfilaments and microtubules connects the nucleus to the location of protein production, the endoplasmic reticulum. Tubules stretch throughout the cell generating components and molecules, based on the cell's specialization (Sloan, 2013).

Chromosomes are often referred to as homologous chromosomes. They are located in practically every cell's nucleus. They have six lengthy strands of DNA that split into 46 individual molecules that couple together to form two chromosomes. Each chromosome has 23 molecules. To create a functioning DNA molecule, it is joined with other proteins to make the chromatins, a tight arrangement of dense fibres strands (Mitchison, 1966).

Every one of the six DNA strands coils over tiny protein complexes called Histones generated by the ER. These combine to produce nucleosome-like structures. DNA molecules carry a negative charge, which is balanced by the positive ions of histones. DNA that is not in use has been coiled and reserved for future usage (Dangl, 2007).

Chromatins are categorized into 2 kinds: type I chromatins and type II chromatins:

- **Euchromatin:** This is the functional section of DNA that's also utilized for RNA transcription, which results in the production of cellular protein necessary for cell growth and differentiation.
- **Heterochromatin:** Heterochromatin is the inert portion of DNA that contains compacted and compressed DNA which is not expressed.

## 7.17. PEROXISOMES OF THE PLANT CELL

These are extremely dynamic tiny organisms with one membrane harboring enzymes that produce hydrogen peroxide. They have critical functions in first and second metabolisms, as well as in the regulation of photorespiration and cell growth in response to biotic as well as abiotic stress (Corpas et al., 2001).

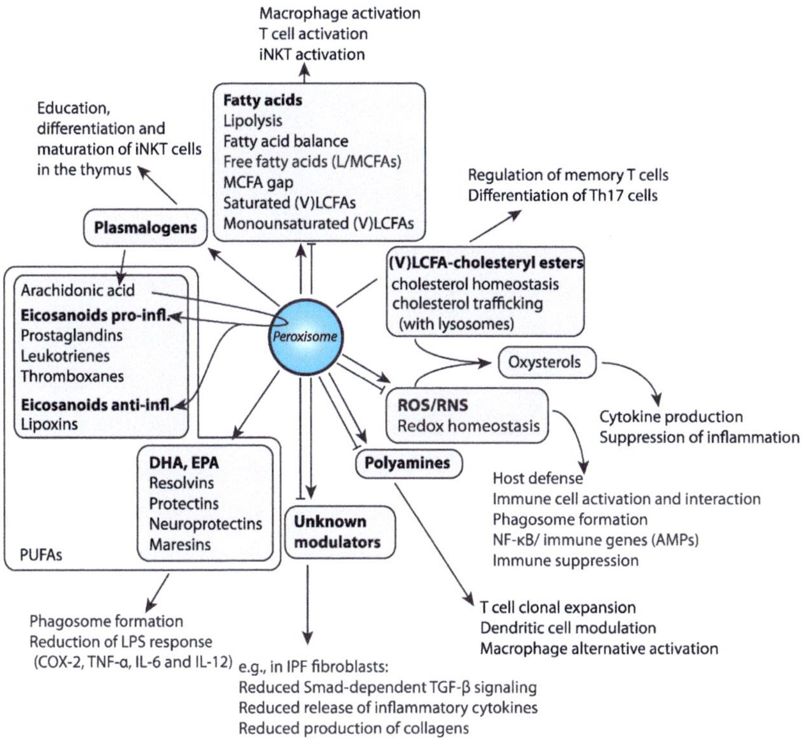


**Figure 7.16.** Peroxisomes are vital organelles.

**Source:** Di Cara F, Andreoletti P, Trompier D, Vejux A, Bülow MH, Sellin J, Lizard G, Cherkaoui-Malki M, Savary S. Peroxisomes in Immune Response and Inflammation. International Journal of Molecular Sciences. 2019; 20(16):3877. https://doi.org/10.3390/ijms20163877

## 7.18. LYSOSOMES IN PLANT CELL

Lysosomes' existence in plants has often been contested, with small evidence for their structural existence. In plants, it is hypothesized as lysosomes develop partly into vacuoles as well as partly into Golgi bodies, that execute the activities assigned to lysosomes (Ueda et al., 2010). Apart from in mammals, wherein lysosomes contain hydrolytic and digesting enzymes for degrading hazardous chemicals and eliminating them from the cell, correspondingly, in plants, such enzymes are united in vacuoles and Golgi bodies. Incomplete differentiating has been compared to the several processes that lead to the creation of Golgi bodies from the endoplasmic reticulum, in which a brief period of lysosomal exudation occurs well beforehand Golgi bodies are completely produced (Matile, 1969).

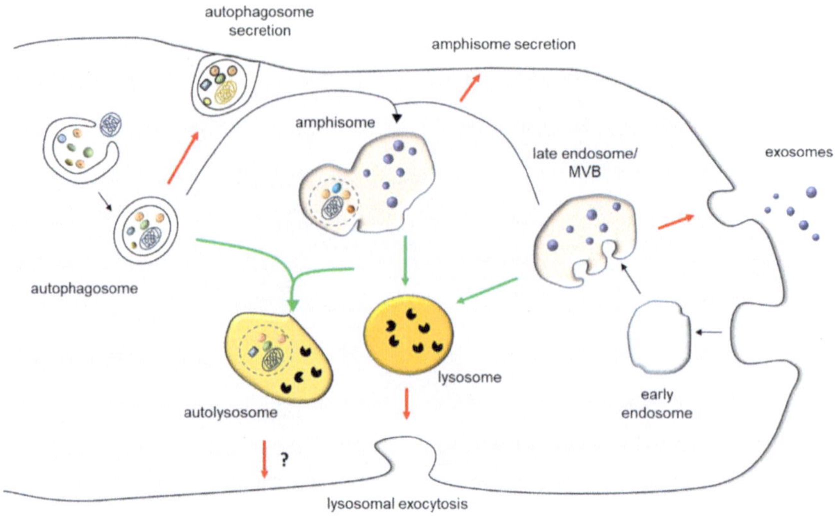


**Figure 7.17.** The biology of Lysosomes.

**Source:** Buratta S, Tancini B, Sagini K, Delo F, Chiaradia E, Urbanelli L, Emiliani C. Lysosomal Exocytosis, Exosome Release and Secretory Autophagy: The Autophagic- and Endo-Lysosomal Systems Go Extracellular. International Journal of Molecular Sciences. 2020; 21(7):2576. https://doi.org/10.3390/ijms21072576

## REFERENCES

1. Akkerman, M., Overdijk, E. J., Schel, J. H., Emons, A. M. C., & Ketelaar, T. (2011). Golgi body motility in the plant cell cortex correlates with actin cytoskeleton organization. Plant and Cell Physiology, 52(10), 1844-1855.
2. Andre, C., Levy, A., & Walbot, V. (1992). Small repeated sequences and the structure of plant mitochondrial genomes. Trends in genetics, 8(4), 128-132.
3. Baluška, F., & Hasenstein, K. H. (1997). Root cytoskeleton: its role in perception of and response to gravity. Planta, 203(1), pp. 69-78.
4. Bock, R. (2001). Transgenic plastids in basic research and plant biotechnology. Journal of molecular biology, 312(3), 425-438.
5. Brandizzi, F. (2021). Maintaining the structural and functional homeostasis of the plant endoplasmic reticulum. Developmental Cell, 56(7), 919-932.
6. Brewin, N. J. (2004). Plant cell wall remodelling in the Rhizobium–legume symbiosis. Critical Reviews in Plant Sciences, 23(4), 293-316.
7. Brunkard, J. O., Runkel, A. M., & Zambryski, P. C. (2015). Chloroplasts extend stromules independently and in response to internal redox signals. Proceedings of the National Academy of Sciences, 112(32), 10044-10049.
8. Byrne, M. E. (2009). A role for the ribosome in development. Trends in plant science, 14(9), 512-519.
9. Cairns, E., Gschwender, H. H., Primke, M., Yamakawa, M., Traub, P., & Schweiger, H. G. (1978). Translation of animal virus RNA in the cytoplasm of a plant cell. Proceedings of the National Academy of Sciences, 75(11), 5557-5559.
10. Camara, B., Hugueney, P., Bouvier, F., Kuntz, M., & Monéger, R. (1995). Biochemistry and molecular biology of chromoplast development. International review of cytology, 163(1), 175-247.
11. Caredda, S., Devaux, P., Sangwan, R. S., & Clément, C. (1999). Differential development of plastids during microspore embryogenesis in barley. Protoplasma, 208(1), 248-256.
12. Chen, J., Doyle, C., Qi, X., & Zheng, H. (2012). The Endoplasmic Reticulum: A Social Network in Plant Cells F. Journal of Integrative Plant Biology, 54(11), 840-850.

13. Cilia, M. L., & Jackson, D. (2004). Plasmodesmata form and function. Current opinion in cell biology, 16(5), 500-506.
14. Corpas, F. J., Barroso, J. B., & del Río, L. A. (2001). Peroxisomes as a source of reactive oxygen species and nitric oxide signal molecules in plant cells. Trends in plant science, 6(4), 145-150.
15. Cosgrove, D. J. (2005). Growth of the plant cell wall. Nature reviews molecular cell biology, 6(11), 850-861.
16. Cosgrove, D. J. (2016). Plant cell wall extensibility: connecting plant cell growth with cell wall structure, mechanics, and the action of wall-modifying enzymes. Journal of experimental botany, 67(2), 463-476.
17. Costes, C., & Vartapetian, B. B. (1978). Plant grown in a vacuum: the ultrastructure and functions of mitochondria. Plant Science Letters, 11(2), 115-119.
18. Dangl, J. L. (2007). Nibbling at the plant cell nucleus. Science, 315(5815), 1088-1089.
19. Darley, C. P., Forrester, A. M., & McQueen-Mason, S. J. (2001). The molecular basis of plant cell wall extension. Plant Cell Walls, Vol. 1, 179-195.
20. DaSilva, L. L., Snapp, E. L., Denecke, J., Lippincott-Schwartz, J., Hawes, C., & Brandizzi, F. (2004). Endoplasmic reticulum export sites and Golgi bodies behave as single mobile secretory units in plant cells. The Plant Cell, 16(7), 1753-1771.
21. Davies, E., Fillingham, B. D., & Abe, S. (1996). The plant cytoskeleton. The Cytoskeleton: A Multi-Volume Treatise, 3(1), 405-449.
22. Deng, Y., Srivastava, R., & Howell, S. H. (2013). Endoplasmic reticulum (ER) stress response and its physiological roles in plants. International journal of molecular sciences, 14(4), 8188-8212.
23. Deslandes, L., & Rivas, S. (2011). The plant cell nucleus: a true arena for the fight between plants and pathogens. Plant Signaling & Behavior, 6(1), 42-48.
24. Di Sano, F., Bernardoni, P., & Piacentini, M. (2012). The reticulons: guardians of the structure and function of the endoplasmic reticulum. Experimental cell research, 318(11), 1201-1207.
25. Fenech, E. J., Ben-Dor, S., & Schuldiner, M. (2020). Double the Fun, Double the Trouble: Paralogs and Homologs Functioning in the Endoplasmic Reticulum. Annual Review of Biochemistry, 89(1), 637-666.

26. Germain, M., Mathai, J. P., & Shore, G. C. (2002). BH-3-only BIK functions at the endoplasmic reticulum to stimulate cytochrome c release from mitochondria. Journal of Biological Chemistry, 277(20), 18053-18060.
27. Goldschmidt-Clermont, M. (1997). Coordination of nuclear and chloroplast gene expression in plant cells. International review of cytology, 177(1), 115-180.
28. Gonzalez, M. E., & Barrett, D. M. (2010). Thermal, high pressure, and electric field processing effects on plant cell membrane integrity and relevance to fruit and vegetable quality. Journal of Food Science, 75(7), pp. 121-130.
29. Gunning, B. E., & Steer, M. W. (1996). Plant cell biology: structure and function. Jones & Bartlett Learning, (Vol. 1, pp. 2-9).
30. Hamann, T. (2015). The plant cell wall integrity maintenance mechanism–a case study of a cell wall plasma membrane signaling network. Phytochemistry, Vol. 112, 100-109.
31. Hammani, K., & Giegé, P. (2014). RNA metabolism in plant mitochondria. Trends in plant science, 19(6), 380-389.
32. Hawes, C., Osterrieder, A., Hummel, E., & Sparkes, I. (2008). The plant ER–Golgi interface. Traffic, 9(10), 1571-1580.
33. Houston, K., Tucker, M. R., Chowdhury, J., Shirley, N., & Little, A. (2016). The plant cell wall: a complex and dynamic structure as revealed by the responses of genes under stress conditions. Frontiers in plant science, 7(1), pp. 984.
34. Kaiserli, E., & Jenkins, G. I. (2007). UV-B promotes rapid nuclear translocation of the Arabidopsis UV-B–specific signaling component UVR8 and activates its function in the nucleus. The Plant Cell, 19(8), 2662-2673.
35. Keegstra, K. (2010). Plant cell walls. Plant Physiology, 154(2), 483-486.
36. Kiss, J. Z., Roberts, E. M., Brown, R. M., & Triemer, R. E. (1988). X-ray and dissolution studies of paramylon storage granules fromEuglena. Protoplasma, 146(2), 150-156.
37. Klimecka, M., & Muszyńska, G. (2007). Structure and functions of plant calcium-dependent protein kinases. Acta Biochimica Polonica, 54(2), 219-233.
38. Knox, J. P., & Benitez-Alfonso, Y. (2014). Roles and regulation of

plant cell walls surrounding plasmodesmata. Current opinion in plant biology, 22(1), 93-100.

39. Kost, B., Mathur, J., & Chua, N. H. (1999). Cytoskeleton in plant development. Current opinion in plant biology, 2(6), 462-470.
40. Lacroix, B., Li, J., Tzfira, T., & Citovsky, V. (2006). Will you let me use your nucleus? How Agrobacterium gets its T-DNA expressed in the host plant cell. Canadian Journal of physiology and pharmacology, 84(3-4), 333-345.
41. Lam, E., Kato, N., & Lawton, M. (2001). Programmed cell death, mitochondria and the plant hypersensitive response. Nature, 411(6839), 848-853.
42. Levy, A., Zheng, J. Y., & Lazarowitz, S. G. (2015). Synaptotagmin SYTA forms ER-plasma membrane junctions that are recruited to plasmodesmata for plant virus movement. Current Biology, 25(15), 2018-2025.
43. Longo, G. P., Pedretti, M., Rossi, G., & Longo, C. P. (1979). Effect of benzyladenine on the development of plastids and microbodies in excised watermelon cotyledons. Planta, 145(3), 209-217.
44. Lopez-Juez, E., & Pyke, K. A. (2004). Plastids unleashed: their development and their integration in plant development. International Journal of Developmental Biology, 49(5-6), 557-577.
45. Mackenzie, S., & McIntosh, L. (1999). Higher plant mitochondria. The Plant Cell, 11(4), 571-585.
46. Martinez, O., Schmidt, A., Salaméro, J., Hoflack, B., Roa, M., & Goud, B. (1994). The small GTP-binding protein rab6 functions in intra-Golgi transport. The Journal of cell biology, 127(6), 1575-1588.
47. Marty, F. (1999). Plant vacuoles. The Plant Cell, 11(4), 587-599.
48. Matile, P. (1969). Vacuoles as lysosomes of plant cells. Biochemical Journal, 111(5), pp. 26.
49. Matsunaga, S., Katagiri, Y., Nagashima, Y., Sugiyama, T., Hasegawa, J., Hayashi, K., & Sakamoto, T. (2013). New insights into the dynamics of plant cell nuclei and chromosomes. International Review of Cell and Molecular Biology, 305, 253-301.
50. McFadden, G. I. (2001). Chloroplast origin and integration. Plant Physiology, 125(1), 50-53.
51. Mitchison, J. M. (1966). Some functions of the nucleus. International Review of Cytology, Vol. 19, 97-110.

52. Mollenhauer, H. H., Whaley, W. G., & Leech, J. H. (1961). A function of the Golgi apparatus in outer rootcap cells. Journal of ultrastructure research, 5(2), 193-200.

53. Northcote, D. H., & Pickett-Heaps, J. D. (1966). A function of the Golgi apparatus in polysaccharide synthesis and transport in the root-cap cells of wheat. Biochemical Journal, 98(1), pp. 159.

54. NOVER, L., MUNSCHE, D., NEUMANN, D., OHME, K., & SCHARF, K. D. (1986). Control of ribosome biosynthesis in plant cell cultures under heat-shock conditions: Ribosomal RNA. European journal of biochemistry, 160(2), 297-304.

55. Ochoa-Villarreal, M., Aispuro-Hernández, E., Vargas-Arispuro, I., & Martínez-Téllez, M. Á. (2012). Plant cell wall polymers: function, structure and biological activity of their derivatives. Polymerization, (Vol. 4, pp. 63-86).

56. Ozcan, L., & Tabas, I. (2012). Role of endoplasmic reticulum stress in metabolic disease and other disorders. Annual review of medicine, 63(1), 317-328.

57. Petrova, V., Pearson, C. S., Ching, J., Tribble, J. R., Solano, A. G., Yang, Y., ... & Fawcett, J. W. (2020). Protrudin functions from the endoplasmic reticulum to support axon regeneration in the adult CNS. Nature communications, 11(1), 1-15.

58. Pozueta-Romero, J., Rafia, F., Houlne, G., Cheniclet, C., Carde, J. P., Schantz, M. L., & Schantz, R. (1997). A ubiquitous plant housekeeping gene, PAP, encodes a major protein component of bell pepper chromoplasts. Plant Physiology, 115(3), 1185-1194.

59. Robson, C. A., & Vanlerberghe, G. C. (2002). Transgenic plant cells lacking mitochondrial alternative oxidase have increased susceptibility to mitochondria-dependent and-independent pathways of programmed cell death. Plant Physiology, 129(4), 1908-1920.

60. Sadali, N. M., Sowden, R. G., Ling, Q., & Jarvis, R. P. (2019). Differentiation of chromoplasts and other plastids in plants. Plant cell reports, 38(7), 803-818.

61. Sakai, A., Takano, H., & Kuroiwa, T. (2004). Organelle nuclei in higher plants: structure, composition, function, and evolution. International review of cytology, 238(1), 59-118.

62. Scharf, K. D., & Nover, L. (1987). Control of ribosome biosynthesis in plant cell cultures under heat shock conditions. II. Ribosomal

proteins. Biochimica et Biophysica Acta (BBA)-Gene Structure and Expression, 909(1), 44-57.

63. Schindler, M., Meiners, S., & Cheresh, D. A. (1989). RGD-dependent linkage between plant cell wall and plasma membrane: consequences for growth. The Journal of cell biology, 108(5), 1955-1965.
64. Schmidt, S. M., & Panstruga, R. (2007). Cytoskeleton functions in plant–microbe interactions. Physiological and Molecular Plant Pathology, 71(4-6), 135-148.
65. Schnepf, E. (1980). Types of plastids: their development and interconversions. Chloroplasts, (Vol. 1, pp. 1-27).
66. Sevilem, I., Miyashima, S., & Helariutta, Y. (2013). Cell-to-cell communication via plasmodesmata in vascular plants. Cell adhesion & migration, 7(1), 27-32.
67. Shibaoka, H., & Nagai, R. (1994). The plant cytoskeleton. Current opinion in cell biology, 6(1), 10-15.
68. Showalter, A. M. (1993). Structure and function of plant cell wall proteins. The Plant Cell, 5(1), 1-9.
69. Showalter, A. M. (1993). Structure and function of plant cell wall proteins. The Plant Cell, 5(1), pp. 1-9.
70. Sloan, D. B. (2013). One ring to rule them all? Genome sequencing provides new insights into the 'master circle'model of plant mitochondrial DNA structure. New Phytologist, 200(4), 978-985.
71. Streward, F. C. (1942). The cytoplasm of the plant cell. Nature, 149(3783), 484-485.
72. Tan, X., Li, K., Wang, Z., Zhu, K., Tan, X., & Cao, J. (2019). A review of plant vacuoles: formation, located proteins, and functions. Plants, 8(9), 327.
73. Thomson, W. W., & Whatley, J. M. (1980). Development of nongreen plastids. Annual Review of Plant Physiology, 31(1), 375-394.
74. Turner, G., Gershenzon, J., Nielson, E. E., Froehlich, J. E., & Croteau, R. (1999). Limonene synthase, the enzyme responsible for monoterpene biosynthesis in peppermint, is localized to leucoplasts of oil gland secretory cells. Plant Physiology, 120(3), 879-886.
75. Ueda, H., Yokota, E., Kutsuna, N., Shimada, T., Tamura, K., Shimmen, T., ... & Hara-Nishimura, I. (2010). Myosin-dependent endoplasmic reticulum motility and F-actin organization in plant cells. Proceedings of the National Academy of Sciences, 107(15), 6894-6899.

76. Voigt, C., Oster, U., Börnke, F., Jahns, P., Dietz, K. J., Leister, D., & Kleine, T. (2010). In-depth analysis of the distinctive effects of norflurazon implies that tetrapyrrole biosynthesis, organellar gene expression and ABA cooperate in the GUN-type of plastid signalling. Physiologia Plantarum, 138(4), 503-519.

77. Wan, S., & Jiang, L. (2016). Endoplasmic reticulum (ER) stress and the unfolded protein response (UPR) in plants. Protoplasma, 253(3), 753-764.

78. Warner, J. R., & McIntosh, K. B. (2009). How common are extraribosomal functions of ribosomal proteins?. Molecular cell, 34(1), 3-11.

79. Wasteneys, G. O., & Yang, Z. (2004). New views on the plant cytoskeleton. Plant Physiology, 136(4), 3884-3891.

80. Weier, E. (1938). The structure of the chloroplast. The Botanical Review, 4(9), 497-530.

81. Weis, B. L., Kovacevic, J., Missbach, S., & Schleiff, E. (2015). Plant-specific features of ribosome biogenesis. Trends in Plant Science, 20(11), 729-740.

82. York, W. S., Darvill, A. G., McNeil, M., Stevenson, T. T., & Albersheim, P. (1986). Isolation and characterization of plant cell walls and cell wall components. In Methods in enzymology (Vol. 118, pp. 3-40). Academic Press.

83. Zambryski, P. (2004). Cell-to-cell transport of proteins and fluorescent tracers via plasmodesmata during plant development. The Journal of cell biology, 164(2), 165.

84. Zavaleta-Mancera, H. A., Thomas, B. J., Thomas, H., & Scott, I. M. (1999). Regreening of senescent Nicotiana leaves: II. Redifferentiation of plastids. Journal of Experimental Botany, 50(340), 1683-1689.

CHAPTER 8

# STRUCTURE AND FUNCTIONS OF ANIMAL CELL

CONTENTS

## 8.1. INTRODUCTION

The primary functional unit of life in animals is the animal cell. It's also the most fundamental unit of reproduction. When microscopy had been established in the 17th century, animal cells were 1st examined. Robert Hooke, an English natural philosopher, was the 1st to identify microscopic holes, subsequently called cells, using plant cork specimens. A Dutch scientist, Anton van Leeuwenhoek, was also able to view cells underneath the microscope. He was the 1st to characterize red blood cells, as well as single-celled species like protozoa and prokaryotic cells (Goldberg, 1993; Minogue et al., 2006).

A eukaryotic cell that has no cell wall and is surrounded via a plasma membrane is known as an animal cell. The cell membrane encloses the cell organelles, such as the nucleus. Plant cells have a cell wall and animal cells have no cell wall.

Animals are a varied collection of living entities that account for around 3/4 of all species in the world. Each of such systems is increased by its constituent components in the body's capacity to move, respond to stimuli, adjust to the variations in the environment, and adjust to varied forms of nutrition, protective factors, and reproduction. Animals, on the other hand, are unable to produce their food; therefore they must rely on plants in some way (Nilsen, 2007; Sakai et al., 2010).

All living things are made up of cells that make up their body structure. Some of these living things are single-celled (unicellular) and other organisms are made up of more than one cell (Multicellular).

A cell is the smallest (microscopic) structural-functional unit of life of an organism. The cells that constitute an animal are called Animal cells and those that constitute plants are known as plant cells.

The cell wall is a protective membrane that surrounds many cells and provides them with its structure and stiffness. The cell wall is a defensive membrane that surrounds many cells and gives them their structure and stiffness.

Animal cells don't have a rigid cell wall; therefore they can generate a wide range of tissues, cell types, and organs. Muscles and nerves are built up of specific cells that plant cells can't develop to make, which allows muscle and nerve cells to move (Barritt, 1999; Gonzalez et al., 2012).

## 8.2. STRUCTURE OF ANIMAL CELL

Animal cells are of different sizes and shapes, ranging from certain mm to μm. The ostrich egg, with a 5-inch diameter and a weight of 1.2-1.4 kilogram, is the biggest animal cell, whereas neurons, with a diameter of around one hundred microns, are the smallest (Clément, 2007; Threadgold, 2017).

Plant cells are bigger than animal cells and, since animal cells have no cell wall, are more amorphous, taking on a variety of shapes. Certain cells are concave, rectangular, and spherical, and others are oval, round, flattened or rod-shaped. The absence of a cell wall is the cause of this. Note that because the majority of the cells are tiny, their anatomy may be studied with a microscope.

However, because both animal and plant cells arose from eukaryotic cells, they share other cell components (Watson & Morris, 1987; Conduit et al., 2015).

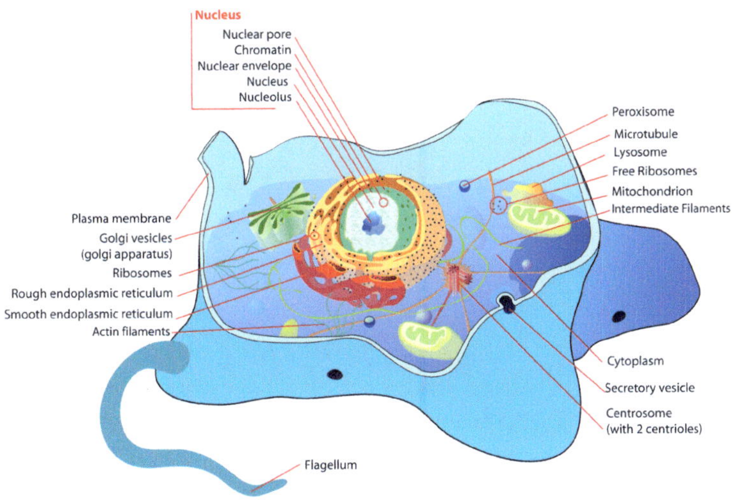


**Figure 8.1.** Animal Cell Structure.

**Source:** https://commons.wikimedia.org/wiki/File:Animal_cell_structure_en.svg

Animal cells, as previously stated, are eukaryotic cells having a membrane-bounded nucleus. As a result, its genetic material is contained in the nucleus in the form of DNA. Inside the plasma membrane, they also contain different structural organelles that conduct different particular jobs for the appropriate function of the cell and to sustain the body's regular systems (Ambros, 2004; Di Leva et al., 2008). The animal cell is composed of numerous structural organelles that are encased in the cell membrane and allow it to work effectively, triggering methods that are beneficial to the animal. All cells enable an animal to recreate, move, react to stimuli, absorb and metabolize food, etc. Usually, it is the united work of all animal cells that allows the body to operate normally (Paoletti et al., 1996).

The following portion describes the organelles found in animal cells:

## 8.3. PLASMA MEMBRANE (CELL MEMBRANE)

Animal cells are surrounded by a protein-membrane sheet which is semi-permeable.

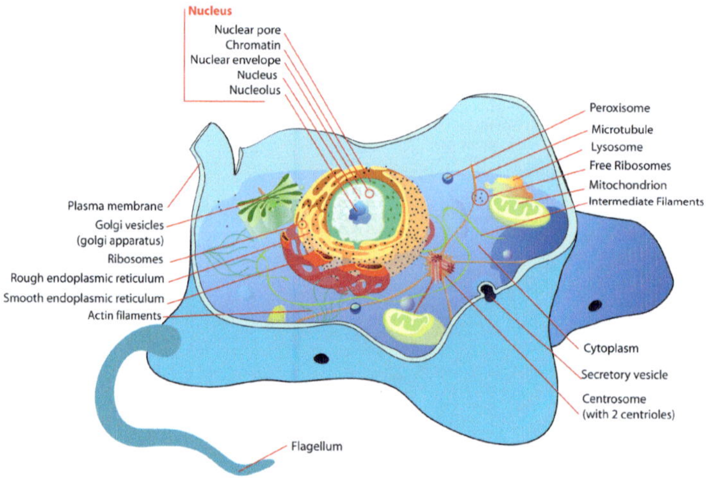


**Figure 8.2.** Plasma Membrane and the cell.

**Source:** https://commons.wikimedia.org/wiki/File:Animal_cell_structure_sv.svg

### 8.3.1. Structure and Functions of Plasma membrane (Cell membrane)

An animal cell's cell membrane is a lipid bilayer having proteins contained in it. The cell membrane's structural architecture allows for selective permeability. Some chemicals won't be capable to get into the cell. Tiny non-polar molecules can be capable of passing through with comfort. Polar molecules, on the other hand, may not, necessitating the use of transporters like membrane proteins. An animal cell's cell membrane has been the sole structure that protects it. Despite the absence of a cell wall, the animal cell's cell membrane includes cholesterol, which offers structural support and stability. Moreover, animal cells are fluid instead of rigid due to the existence of cholesterol and contain no cell walls, and hence have the flexibility to move (Ishikawa et al., 1997; Jimenez et al., 2016).

It encompasses and safeguards the contents of the cell. It also controls the chemicals that flow through the plasma membrane to enter and exit the cell. As a result, it maintains homeostasis. The proteins play an important role in moving materials throughout the membrane. Carbohydrates, which embellish both lipids and proteins and enable cells to identify one another, enable communication of cells (Nicolson, 1974).

## 8.4. NUCLEUS

This specialized organelle, the nucleus, acts as the cell's data and organizational hub. Two essential roles are served by this cellular structure. Among its functions are DNA storage and coordination, as well as intermediate metabolism, development, synthesis of protein, and cellular reproduction (Wu, 1995; Butler, 2005).

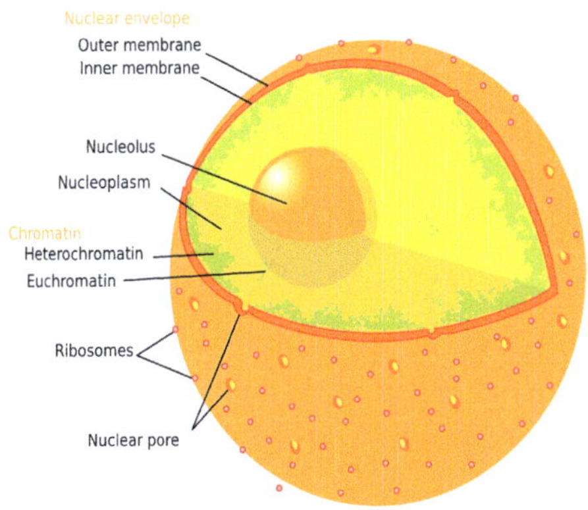


**Figure 8.3.** Animal Cell's Nucleus.

**Source:** https://commons.wikimedia.org/wiki/File:Diagram_human_cell_nucleus.svg

Only advanced organisms, referred to as eukaryotes, contain cells with a nucleus. Usually, each cell has a single nucleus, although certain outliers exist, like slime molds and the Siphonales group of algae. Prokaryotes, such as cyanobacteria and bacteria, are simpler one-celled creatures that lack a nucleus. In such species, the cytoplasm contains all of the cell's data and administrative activities (Fededa, & Gerlich, 2012).

The spherical nucleus comprises around 10% of the volume of a cell, making it the cell's most visible characteristic. The majority of nuclear material is composed of chromatin, the unstructured state of the cell's DNA that would assemble into chromosomes throughout mitosis or cellular division. The nucleolus is also located within the nucleus; it is an organelle that synthesizes protein-producing macromolecular assemblies known as ribosomes (Watson, 1959).

The nuclear envelope, a double-layered membrane, separates the composition of the nucleus from the cytoplasm of the cell. The membrane is punctured by openings known as nuclear pores that enable particular sizes and types of molecules to move between the cytoplasm and the nucleus. Additionally, it has been connected to a network of tubules termed the endoplasmic reticulum, which is responsible for protein production. Such tubules run the length of the cell and synthesize the biochemical products for which a specific cell type is genetically programmed (Maxfield, 2002; Prenner et al., 2005).

**Chromatin/Chromosomes:** Each human cell has over six feet of DNA packed within its nucleus, which is separated into forty-six distinct molecules, one per chromosome and every measuring around 1.5 inches long. All material must be packed into a minuscule cell nucleus, which is a remarkable feat of packing. DNA cannot be jammed into the nucleus like a ball of thread to operate. Rather than, it is joined with proteins and structured into a specific, tight structure known as chromatin, which is a thick string-like fibre.

Every DNA strand wraps around groups of tiny protein molecules referred to as histones to create nucleosomes, which are a sequence of bead-like structures joined by the DNA strand. Uncondensed chromatin appears like "beads on a thread" under the microscope (Harder et al., 1998; Serrano, 1989).

The nucleosome string, which has already been compressed via a factor of 6, is next coiled into a tighter structure, squishing the DNA via a factor of forty. DNA compression and structuring serve several purposes. The DNA's total -ve charge is neutralized via histone molecules' +ve charge, the DNA takes up considerably less space, and inactive DNA may be folded into inaccessible areas until it has been required.

Chromatin is divided into two categories. Euchromatin is the genetically active component of the cell, and it is responsible for transcribing RNA into proteins that are needed for cell function and growth. Heterochromatin contains inactive DNA and is the most condensed component of chromatin since it is not utilized (Luna & Hitt, 1992).

Inside the nucleus, chromatin fibres take on several shapes during a cell's life. The chromatin is scattered throughout the nucleus in what looks to be a tangle of fibres during interphase when the cell is carrying out its usual duties. This exposes the euchromatin, allowing the transcription process to take place.

The chromatin modifies substantially whenever the cell approaches metaphase and begins to divide. To begin, all of the chromatin strands duplicate themselves via the DNA replication mechanism. The cells are then squeezed ten thousand times more than at interphase, resulting in specialized structures for reproduction known as chromosomes. The chromosomes split when the cell splits into 2 cells, providing every cell with a full copy of the genetic data stored in the chromatin (Uemura et al., 2006; Ingólfsson et al., 2014).

### 8.4.1. Parts of the Nucleus

The **nucleolus** is a membrane-free organelle inside the nucleus which produces ribosomes, which are the cell's protein-producing structures. The nucleolus appears as a huge black patch inside the nucleus under a microscope. A nucleus can have 4 nucleoli; however, the quantity of nucleoli in every species is set. As chromosomes are gathered together into nucleolus organizing areas after a cellular division, a nucleolus is produced. The nucleolus vanishes throughout the cellular division. Certain studies imply that the nucleolus is implicated in cellular aging and, as a result, can influence an organism's aging (Ravikumar et al., 2010; Bernardino et al., 2016).

**Nuclear Envelope:** in a cell's lifecycle, the nuclear envelope has is a double-layered membrane that keeps the composition of its nucleus contained for the majority of the time. In the perinuclear region, the area between the layers is referred to as the RER, and it seems to be connected to it. The envelope that is pierced contains microscopic holes known as nuclear pores, which allow for the passage of information. These pores govern the movement of molecules between the nucleus and the cytoplasm, allowing certain molecules to diffuse while prohibiting others from doing so. The inner side of the nucleus is lined by a protein coating known as the nuclear lamina that interacts with chromatin as well as other nuclear constituents. Throughout mitosis, the nuclear envelope decomposes and afterward reforms as the 2 cells are formed and the chromatin proceeds to unwind and scatter, resulting in the development of 2 new cells (Steinman et al., 1983).

**Nuclear Pores:** Nuclear pores are openings in the nuclear envelope that allow for the passage of atoms. Such holes govern the movement of molecules between the nucleus and the cytoplasm, allowing certain molecules to cross the membrane while prohibiting others from doing so. The entry of basic components for the construction of RNA and DNA and molecules which supply energy for the construction of genetic information, into the nucleus is permitted (Nakatsu et al., 2012).

They are completely accessible to tiny molecules with sizes as small as those of proteins, yet they act as a barrier, stopping most big molecules from entering the nucleus altogether. A small number of bigger proteins, like histones, are permitted entry into the nucleus. Every pore has been accompanied via a complicated protein structure known as the nuclear pore complex that is thought to choose big molecules for entry into the nucleus by interacting with them (Clegg, 1984).

## 8.5. CYTOPLASM

The cytoplasm is the composition of a cell outside of the nucleus, and it has been contained inside the cell membrane. It has a gel-like texture and is colorless. Water makes up the bulk of the cytoplasm, however, it also incorporates salts, enzymes, organelles, and diverse chemical substances.

Organelles and cell molecules are sustained and suspended in the cytoplasm.

The production of protein, the earliest phase of cell respiration, meiosis and mitosis, are all cellular functions that take place in the cytoplasm. The cytoplasm facilitates the transport of components inside the cell, such as hormones, as well as the dissolving of cellular waste. The cytoplasm is separated into 2 sections: ectoplasm (ecto-,-plasm) and endoplasm (endo-,-plasm). The organelles have been found in the endoplasm, which is the center portion of the cytoplasm. The ectoplasm is a gel-like outer component of a cell's cytoplasm (Rose et al., 1991; Mitchison et al., 2008).

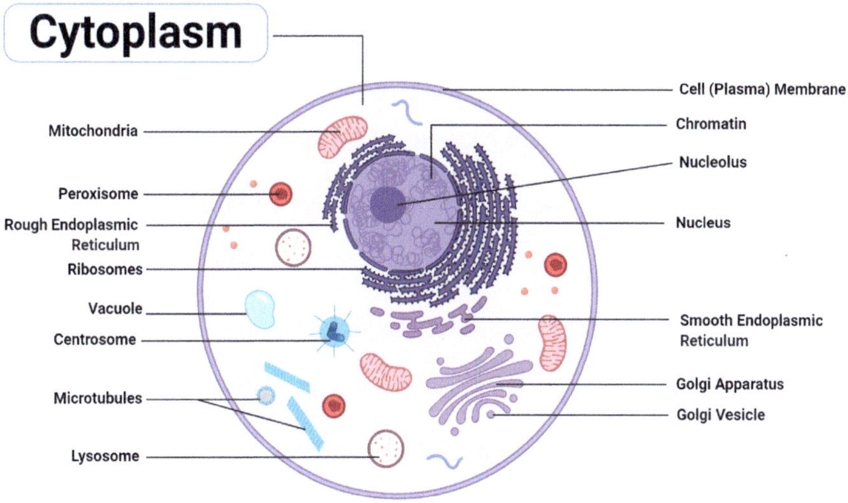


Figure 8.4. The Animal Cell's Cytoplasm

Source: https://slideplayer.com/slide/14715862/

### 8.5.1. Components of Cytoplasm

Archaeans and bacteria are examples of prokaryotic cells that lack a membrane-bound nucleus. The cytoplasm in such cells has been made up of all of the elements of the cell that have been restricted inside the cell membrane. The cytoplasm of eukaryotic cells, like animal and plant cells, is made up of 3 basic elements. The organelles, cytosol, and cytoplasmic inclusions have been the organelles, cytosol, and different granules and particles (Li et al., 2013).

The **cytosol** has been the semi-fluid element of a cell's cytoplasm or liquid media. It is the outer side of the nucleus and the inner side of the cell membrane.

**Organelles** are tiny cell structures that have a variety of activities inside the cell, including the production of energy. Among the organelles found in cells are ribosomes, the nucleus, ribosomes, the chloroplast, the endoplasmic reticulum, the apparatus of Golgi, and the chloroplast apparatus. Additionally, the cytoskeleton, which is a network of fibers that helps the cell maintain its shape and provides support for organelles, is present within the cytoplasm (Koch & Götz, 1972; Moeendarbary et al., 2013).

**Inclusions of Cytoplasmic:** A cytoplasmic inclusion is a collection of particles that have only been suspended in the cytoplasm for a brief amount of time. Inclusions include macrophages and granules, to name a few examples. Exosomes, inclusions of secretory origin, inclusions of nutritive origin, and pigment granules are the three types of inclusions found in the cytoplasm of the cell. Secretory inclusions include enzymes, p proteins, and acids, to name a few examples. Glycogen and lipids are examples of nutritive inclusions. The pigment granule inclusion known as melanin, which is present in skin cells, is an example (Bray & White, 1988).

### 8.5.2. Cytoplasmic Streaming

Cytoplasmic streaming, also known as cyclosis, is the circulation of chemicals inside a cell. Numerous types of cells, particularly amoeba, plant cells, protozoa, and fungus, exhibit cytoplasmic streaming. Many variables can alter cytoplasmic mobility, such as the existence of specific substances, hormones, and variations in light or temperature (Kite, 1913).

Plants use cyclosis to transport chloroplasts to regions where the greatest sunshine is accessible. Photosynthesis has been carried out by chloroplasts, which are organelles of plants that need light to function. Cytoplasmic streaming is employed for movement in protists like amoebae and slime molds. Pseudopodia are temporary extensions of the cytoplasm that have been useful for moving about and catching food. Because the cytoplasm should be dispersed across daughter cells created during meiosis and mitosis, cytoplasmic streaming is also essential for cellular division (Kite, 1915).

## 8.6. MITOCHONDRIA

All eukaryotic cells have such membrane-bound organelles in their cytoplasm. The number of mitochondria in every cell differs significantly based on the cell's job. Erythrocytes, for instance, lack mitochondria, but the muscle and liver cells contain thousands of them (Lenaz et al., 2006; Kirillova et al., 2021).

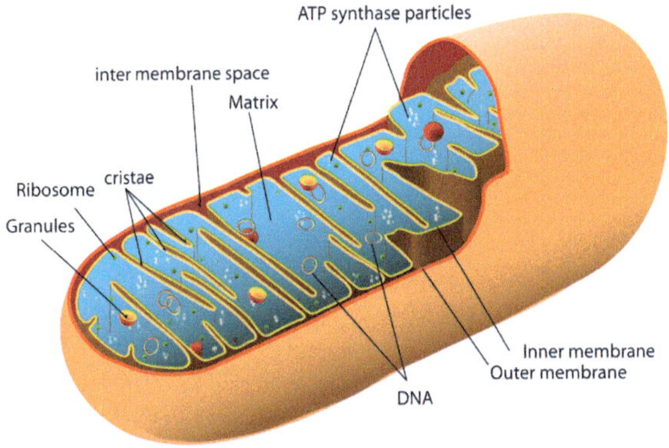


**Figure 8.5.** Animal Cell Mitochondria Structural Interpretation.

**Source:** https://biologydictionary.net/mitochondrion/#foobox-2/0/Animal-mitochondrion-diagram.jpg

## 8.6.1. Structure and Functions of Mitochondria

They are rod-shaped, oval-shaped, or spherically shaped, and range in size from half to ten meters. Mitochondria contain 2 membranes: an external membrane and an internal membrane. In the core bulk, there is a mitochondrial gel-matric. Cristae are folds formed by the membranes (Wang & Youle, 2009).

They are the powerhouses for the cell, creating energy in the form of Adenosine Triphosphate (ATP) by transforming oxygen and nutrients into energy, allowing the cell to fulfill its role while also releasing surplus energy from the cell.

Mitochondria also contain Ca, which helps in cell signaling, cell and physical heat generation, and cell development and death.

The exterior membrane is permeable, enabling tiny molecules to pass through and big molecules to pass via a separate route.

The internal mitochondrial membrane is less porous, enabling extremely tiny molecules to enter the core mass's mitochondrial gel matrix. The mitochondrial DNA and enzymes for the Tricarboxylic Acid (TCA) cycle, also known as the Kreb's Cycle, make up the gel matrix (Lee et al., 1997; Seo et al., 2010).

The TCA cycle depletes nutrients by turning them into by-products that are used by mitochondria to generate energy. Since the membrane twists into folds termed cristae, whereas the protein elements necessary for the major energy generation system cells referred to as the Electron Transport Chain (ETC), occur, such activities occur in the inner membrane. ETC is the body's main resource of ATP generation.

The ETC consists of numerous oxidation-reduction activities that move electrons from the first protein element to the second protein element, generating energy that is then utilized to phosphorylate ADP (adenosine diphosphate) to ATP. The chemiosmotic coupling of oxidative phosphorylation is the name for such a mechanism. Several cellular processes, particularly muscular movement, are powered by such a system, which also boosts overall brain performance.

The cell nucleus provides some, though not all, of the proteins and chemicals that are composed of the mitochondria. The mitochondrial nucleus genome contains thirty-seven genes, thirteen of which create the majority of the ETC elements. Moreover, as mitochondrial DNA lacks a significant DNA repair mechanism used in other nuclear DNAs, it is extremely prone to mutations (Reddy, 2009).

Furthermore, due to the predilection for abnormal generation of free electrons, Reactive Oxygen Species (ROS), also called free radicals, have been generated in the mitochondrion. Antioxidant proteins in the mitochondrion neutralize such electrons. Certain free radicals, meanwhile, may harm mitochondrial DNA (mtDNA).

It is also possible to damage mtDNA through excessive alcohol utilization by oversaturating detoxifying enzymes, resulting in the generation and spillages of strongly reactive electrons into the cytoplasmic membrane and mitochondrial matrix, which combine with some other cell molecules to develop multiple radicals that markedly harm cells (Minibayeva et al., 2012; Dong et al., 2017).

Several organisms acquire their mother's mtDNA. As a result, the mother's cytoplasm is transferred to the fetus in large quantities, whereas the mitochondria from the father's sperm have been damaged. Changes in the mother and father DNA or mother mtDNA produces mitochondrial disorders that are both hereditary and developed. Alzheimer's disease and Parkinson's disease are examples of these disorders. Aging and some malignancies and disorders have been related to mtDNA mutations that accumulate over time.

Because of mutations in the mitochondrial DNA (mtDNA), mitochondria naturally play an important part in the process of programmed cell death (apoptosis) (Borst, 1977).

## 8.7. RIBOSOMES

"The most essential cell organelles are ribosomes, which have been made up of protein and RNA and transform genetic information into amino acids chains".

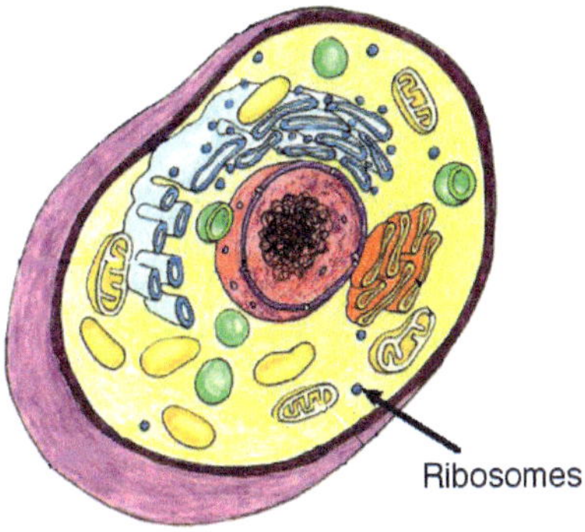


**Figure 8.6.** Ribosome structural depiction.

**Source:** https://joshandkarlasproject.weebly.com/nucleolus-nucleous-ribosomes-and-vacuole.html

This complex molecular instrument, known as a ribosome, is responsible for the production of proteins from amino acids during the process of protein synthesis, also known as translation. Protein synthesis is a fundamental action that all living cells must perform to function properly (Hartley et al., 1996; de la Cruz et al., 2015).

Ribosomes, which are specialized cell organelles found in both eukaryotic and prokaryotic cells, are found in both types of cells. Ribosomes are necessary for the production of proteins throughout the body of a living cell.

Binding to an mRNA and interpreting the data contained by the mRNA's nucleotide pattern are other activities of such cell organelle. They transmit amino acid-containing RNAs to the acceptor site of the ribosome. It binds to tRNA and then adds amino acids to the developing protein chain (Ross & Zarbalis, 2014; Joazeiro, 2019).

### 8.7.1. Ribosomes Structure

A ribosome is a ribonucleoprotein, which means it is made up of protein and RNA. It is made up of 2 parts: the first one is small and the second one is large. The mRNA attaches and is translated in the small subunit, while the amino acids are added in the bigger subunit. Ribonucleic acid and protein are both present in both subunits (Stewart & Thomas, 1994; Caldarola et al., 2009).

Connections among rRNAs inside the first subunit and proteins inside the second subunit link the 2 subunits together. In both animal and plant cells, ribosomes are situated within the cytoplasm.

The following components make up the ribosome structure:

- It can be found in 2 different parts of the cytoplasm.
- The cytoplasm is strewn with them.
- Eukaryotes contain 80S ribosomes, whereas prokaryotes contain 70S.
- RNA makes up around sixty-two percent of ribosomes, whereas proteins make up the balance.
- Bound and free ribosomes have the same structure that is linked to the synthesis of proteins.

### 8.7.2. Ribosomes Function

The primary job of the ribosome is to assemble amino acids into proteins, which are essential for the completion of many cellular processes. The mechanism of DNA transcription results in the synthesis of messenger RNA (mRNA), which is derived from DNA. The nucleus generates messenger RNA (mRNA), which is subsequently transported to the cytoplasm to initiate the protein synthesis processes. Ribosomal subunits are found in the cytoplasm, where they are wrapped around mRNA polymers. The tRNA then goes on to function as a protein synthesis enzyme. It has been shown that proteins created in the cytoplasm are utilized within the cell, but proteins made by bound ribosomes are transported outside the cell (Boon et al., 2001; Magnuson et al., 2012).

## 8.8. ENDOPLASMIC RETICULUM (ER)

Cellular organelle which links from the cytoplasm to the nucleus that's also an uninterrupted folded membrane organelle observed in the cytoplasm and is composed of a slender network of flattened connecting sections that links from the cytoplasm to the nucleus. There have been membranous gaps inside its membranes, which are referred to as cristae spaces, and the membrane folding is referred to as cristae (Smirnov et al., 2011; Gościńska & Topf, 2020).

The RER and SER are 2 forms of ER that differ in their structure and the functions they execute.

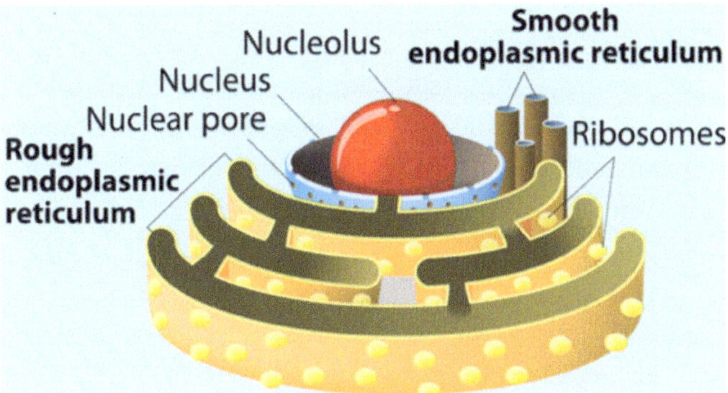


**Figure 8.7.** Endoplasmic Reticulum structural depiction.

**Source:** https://microbenotes.com/endoplasmic-reticulum-er-structure-types-functions/

## 8.8.1. Endoplasmic Reticulum (ER) Functions

Producing, digesting, and delivering proteins for use both inside and outside the cell. This is because it is unswervingly attached to the nuclear membrane, which allows for a passageway between cytoplasm and nucleus.

Because the ER contains greater than 50% of the substance of membrane cells, it provides a wide surface area wherein chemical interactions might occur. They also include enzymes for approximately all of the cell's lipid production, making them the lipid generation site (Lynes & Simmen, 2011; Oakes & Papa, 2015).

The ER is divided into two categories based on its morphological and functional properties: RER and SER.

## 8.8.2. Types of Endoplasmic Reticulum

**Rough Endoplasmic Reticulum (RER):** RER gets its name from the fact that it has a rough texture coated in ribosomes. It is the ribosomes on the RER that are responsible for protein creation, and they have a signaling pattern that leads them to the ER where they will be processed further. The RER is responsible for transporting proteins and lipids into the cristae across the cell. They are then either injected into the cell membrane or transported into the Golgi bodies, depending on the situation (Bhattacharyya & Glick, 2007; Xiang et al., 2017).

**Smooth Endoplasmic Reticulum (SER):** Even though it is located close to the RER, the SER is not related to ribosomes and has a function separate from that of the RER. Its function is to produce lipids (phospholipids and cholesterol), which are subsequently utilized in the formation of a new cell membrane. Cholesterol-producing cell types are related to the production of steroid hormones, according to research. It also assists in the detoxification of the liver after the absorption of medications or toxic chemicals.

The sarcoplasmic reticulum (SER) is a highly specialized variant of the endoplasmic reticulum. Muscle cells rely on it to regulate the concentration of calcium ions in their cytoplasms within acceptable limits (Eizirik et al., 2008; Fu et al., 2012).

## 8.9. GOLGI APPARATUS

Cell organelles are membrane-bound and present in the cytoplasm of eukaryotic cells, adjacent to the ER and close to the nucleus. Golgi bodies are kept together via cytoplasmic microtubules, which are in turn kept together via a protein matrix. Cisternae are pouches that have been flattened and stacked together to form a structure. For animal cells, the number of cisternae can range from four to ten in number. Although certain species, such as single-celled creatures, have roughly sixty cisternae, the majority of them have Golgi bodies (Tabas, 2010; Mao et al., 2019).

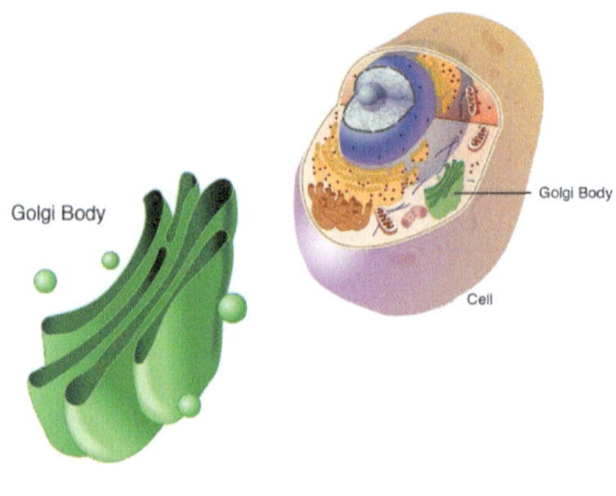


**Figure 8.8.** Golgi Apparatus Structural Depiction.

**Source:** https://www.genome.gov/genetics-glossary/golgi-body

They are divided into 3 basic sections: the cis (Cisternae closest to the Endoplasmic Reticulum), the medial (middle layers of cisternae), and the trans (Cisternae away from the ER).

Animal cells contain certain Golgi bodies (1-2), but plants have hundreds of Golgi bodies.

## 8.9.1. The Golgi Apparatus Performs a variety of Functions

Their primary function is to transport, modify, and pack proteins and lipids into Golgi vesicles for them to be transported to their respective destinations inside the body. In contrast to animal cells, which have one or more Golgi bodies, plants possess a few hundred Golgi bodies.

At the cis and trans faces of the cisternae, the exterior layer is composed of cis and trans-Golgi networks, which are responsible for sorting the lipids and proteins that are gathered at the cis face and expelled by the trans face through the Golgi bodies (Seemann et al., 2002; Volgyi et al., 2015).

It is on the cis face of the joined vesicles that the lipids and proteins are gathered in large groups. Specifically, the vesicular-tubular cluster is a sector of the cell that permits the fused vesicles to travel along the microtubules with relative ease. This portion is located between the Golgi apparatus and the endoplasmic reticulum.

It is believed that the cis Golgi network is produced by the merging of the vesicle-cisternal clusters, which transports lipids and proteins into the cis face cisternae, where they are converted into functional units as they move from the cis to the trans face. Cellular external and intracellular elements are both served by functional units that are carried throughout the cell (Pagano et al., 1991; Murshid & Presley, 2004).

Phosphorylation is used to introduce fatty acids and phosphate groups into cells, as well as to delete monosaccharides, such as by removing mannose moieties from the cis and medial cisternae and introducing galactose into the trans cisternae, among other things.

It is in the trans-Golgi network that the proteins and lipids that have been altered are separated and packed into trans vesicles that carry them to the lysosomes, or rarely to the cell membrane, for exocytosis. In the presence of ligands bound to receptors, fusion and protein release occur, which is assisted by the presence of ligands bound to receptors (Hibbard, 1945; Pagano, 1988).

## 8.10. LYSOSOMES

The primary purpose of such small organelles is to act as digestive chambers for cellular elements which have outlived their usefulness or otherwise have become obsolete. The lysosomes renew the cell's organic matter in this way, which is called autophagy. Cellular waste material, lipids, proteins, carbohydrates, and other macromolecules are broken down into simple chemicals by lysosomes, which have been subsequently transported back into the cytoplasm as novel cell-building components. The lysosomes use roughly forty distinct kinds of hydrolytic enzymes to perform the functions involved with digestion, which are all produced in the ER and altered in the Golgi apparatus. Lysosomes are frequently budded from the Golgi apparatus membrane, but they may also emerge progressively from late endosomes, which have been vesicles that bring components into the cell via endocytosis (Franke et al., 1972; Driouich et al., 1993).

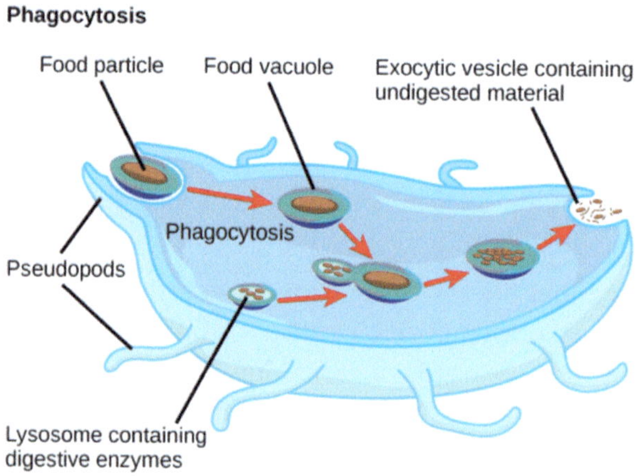


**Figure 8.9.** Structure of Lysosome.

**Source:** https://commons.wikimedia.org/wiki/File:Figure_04_04_04.jpg

Microbodies such as lysosomes are sphere-shaped organelles that are confined within a single-layer membrane; however, their size and form can vary to a certain degree depending on their location within the cell. Lysosomes have strong digestion enzymes that would otherwise cause severe harm if they were allowed to enter the remainder of the cell, which is why this barrier is so important. The cell is further protected from exposure to the biochemical catalysts present in lysosomes because they require an acidic environment to function properly. A pH of around 4.8 is optimal for enzyme activity in the lysosomal matrix, but the neutral environment of the cytoplasm makes most digestive enzymes inactive, thus even in the event of a lysosomal membrane breach, the cell may survive. It is possible to regulate the acidity of the lysosome with the assistance of hydrogen ion pumps, and the organelle can resist self-digestion by glycosylating internal membrane proteins to avoid their breakdown (Satiat-Jeunemaitre & Hawes, 1992).

The identification of lysosomes included the usage of a centrifuge to segregate the different cellular components. As part of his research into the carbohydrate metabolism of liver cells in the mid of 20th century, the Belgian scientist Christian René de Duve discovered that the cells produced an enzyme known as an acid phosphatase in greater quantities as they were subjected to proportionally greater centrifugal stress. According to de Duve's theory, the digesting enzyme had been encased in a certain form of membrane-bound organelle inside the cell, which he termed the lysosome, to explain the phenomena. In photos captured using an electron microscope, after calculating the likely size of the lysosome, he had been capable to recognize the organelle (Seemann et al., 2000).

Many animal cells include lysosomes, although disease-fighting cells, like white blood cells, have the most. It's because, to combat bacteria, viruses, as well as other external invaders, white blood cells should consume extra substance than most other cell types. Numerous illnesses in humans are caused via lysosome enzyme abnormalities which prevent cellular digestion. A genetic mutation, for instance, causes Tay-Sachs disease, which hinders the development of a crucial enzyme that breaks down complicated lipids termed gangliosides. Such lipids build up in the neurological system, causing mental impairment and even death in children. Inflammation and discomfort in arthritis are also linked to lysosome enzyme release (Webster et al., 1978; Carraway & Carraway, 1989).

## 8.11. CYTOSKELETON

Essentially, it is a fibrous web that has been generated by several proteins which include lengthy chains of amino acids. These proteins are present in the cytoplasm of eukaryotic cells, where they function as transporters.

Microtubules, actin filaments, and intermediate filaments are the three types of small filaments that make up the cell (Barak et al., 1980; Mège et al., 2006).

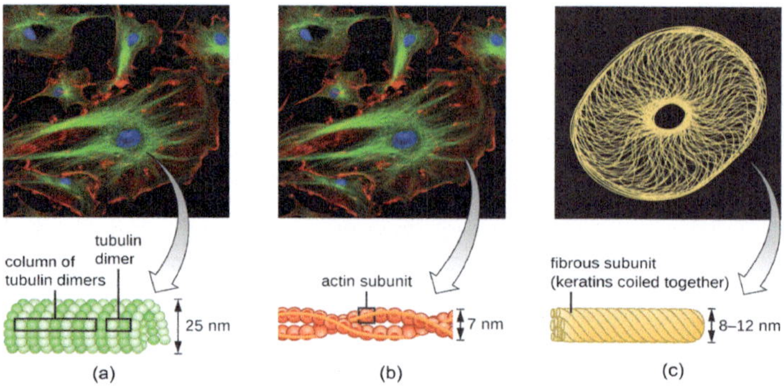


**Figure 8.10.** Structure of the cytoskeleton.

**Source:** https://commons.wikimedia.org/wiki/File:0317_Cytoskeletal_Components.jpg

## 8.11.1. Functions of Cytoskeleton

The cytoskeleton acts to establish a network coordinating the cellular elements and also preserving the cellular structure. The filamentous system network present in the cell's cytoplasm also allowed for regular mobility of the cell as well as its organelles. This also arranges a few of the cellular elements maintaining the cellular structure. It is important for cell mobility and the mobility of various cell organelles in the cytoplasm (Hall, 2009; Zhou et al., 2015).

Among the teeny-tiny filaments are:

**Actin filaments**, also called microfilaments, are a meshwork of parallel fibers that assist the cell motion and mediate specific cellular activity like substrate adhesion and cleavage processes throughout the mitotic cellular division. They alter continuously, assisting the cell motion and intervening in definite cellular activity like substrate adhesion capability and cleavage processes throughout the mitotic cellular division (Ierushalmi & Keren, 2021).

**Microtubules** are lengthy filaments that help to move chromosomes of daughter cells to new developing daughter cells during mitosis.

In comparison to microtubules and actin, intermediate filaments are more persistent filaments. They create the cell's real structure and keep the nucleus in its proper place inside the cell. It also helps the cell's elasticity component to function properly, allowing it to withstand physical stress.

Septin (that arranges filaments) and spectrin are two more proteins that can be introduced to the cell's cytoskeleton (that keeps the cell's structure together by pulling the cell membrane and the intracellular surface of the cell together) (Picone et al., 2010; Sasabe & Machida, 2012).

## 8.12. MICROTUBULES

Microtubules are microscopic empty cylinders formed of the proteins α and β-tubulin that have been components of the cell's cytoskeleton. The cytoskeleton is a system of protein fibers that runs all across the cell, provides its structure, and maintains its organelle. Microtubules, which are around twenty-four nanometers thick, are the large structures in the cytoskeleton. They play a part in cell mobility, cellular division, and substance transfer inside cells (Karsenti & Maro, 1986; Sasabe & Machida, 2012).

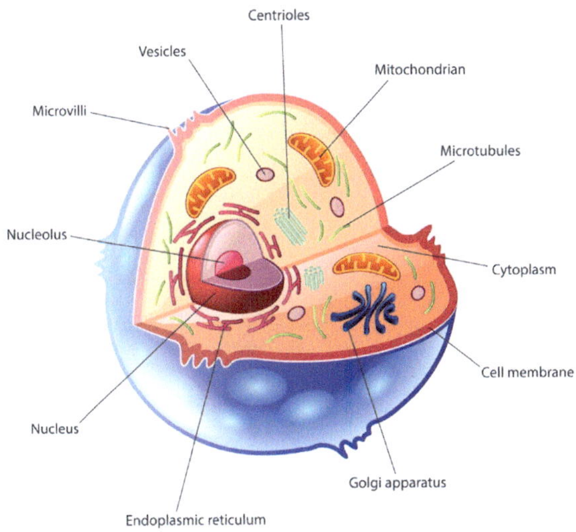


**Figure 8.11.** Microtubules have a unique interior structure.

**Source:** https://biologywise.com/animal-cell-functions

### 8.12.1. Microtubule Structure

Microtubules are empty tubes composed of repetitive protein complexes, namely alpha and beta-tubulin dimers (also referred to as □-tubulin and β-tubulin in written form). Dimers are 2 structures of protein. □-tubulin and β-tubulin connect to create a dimer, and afterwards, numerous units of such dimers link with each other in a chain termed a protofilament, constantly switching alpha and beta. To create a microtubule, 13 protofilaments are arranged in a cylindrical arrangement. Through the adding and removing of dimers, microtubules continually assemble and disassemble. Since its structure is conserved even though the single molecules are continually altering, they have been considered to be in a state of dynamic equilibrium (Jacobs et al., 1988; Akhmanova & Kapitein, 2022).

Microtubules comprise polar molecules including one +ve charged side that develops quickly and the other -ve charged side that grows slowly. In a microtubule, protofilaments align exactly parallel to one other, leaving β subunits uncovered on the +ve side and a subunits uncovered on the -ve side. Polarity enables the microtubule to form and operate appropriately in a precise manner.

Microtubules in animal cells extend outwardly from a microtubule-organizing center (MTOC) termed a centrosome, which is located at the cell's center. Fungi and plants don't have centrosomes; rather, the nuclear envelope of the membrane that surrounds the nucleus is an MTOC (Stephens & Edds, 1976; Vasiliev & Samoylov, 2013).

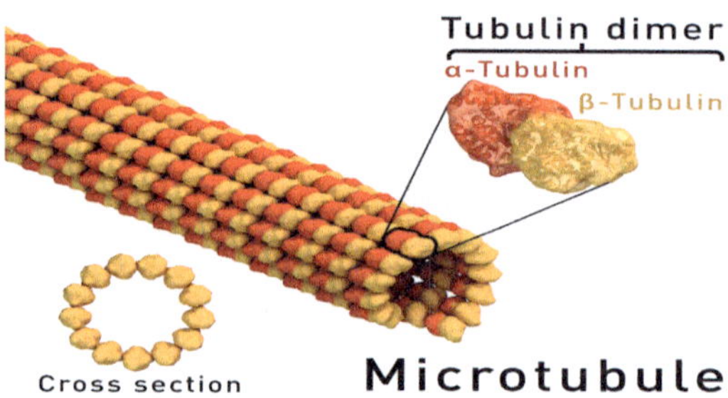


**Figure 8.12.** A microtubule's structure.

**Source:** https://biologydictionary.net/microtubule/

## 8.12.2. Function of Microtubules

### *8.12.2.1. Cell Movement*

Structures including flagella and cilia are made up of microtubules. Cilia are tiny protuberances that protrude from a cell. They are present in the cells that line the trachea in humans, wherein they protect mucus and debris from reaching the lungs. They're also present in the female reproductive system's fallopian cylinders; wherein they assist carry the egg produced from the ovary to the uterus. Flagella are extensions that resemble tails and enable cells to move. Flagella are present in several bacteria, and male sperm also use them to travel about. Microtubules also enable whole cells to "crawl" or move from one location to another via constricting at one side and growing at the other (Tuszynski et al., 1997; Conduit et al., 2015).

#### *8.12.2.2. Cell Division*

Microtubules are essential for the formation of the mitotic spindle, also known as the spindle apparatus. In eukaryotic cells, that structure is generated throughout mitosis. Throughout cellular division, the mitotic spindle arranges and divides chromosomes so that they may be divided into 2 different daughter cells. Microtubule-associated proteins (MAPs), microtubules, and the MTOCs are among its constituents.

Polar, astral, and kinetochore microtubules are 3 kinds of microtubules that help with mitosis. The mitotic spindle is kept in position by astral microtubules that extend from a cell's MTOCs to the cell membrane. Polar microtubules connect 2 MTOCs and aid in chromosomal separation. (Polar microtubules are typically polar; such are only termed polar microtubules.) Kinetochore microtubules bind to chromosomes to aid in their separation; the chromosomes are coupled to the microtubules via protein's kinetochore complex (Lange & Gull, 1996; Boyko et al., 2000).

#### *8.12.2.3.* **Cell Transport**

Microtubules, which have been components of the cytoskeleton, assist in the movement of organelles within the cell's cytoplasm, which includes all the cell's constituents except the nucleus. They also assist the cell's different portions in communicating with one another. Microtubules, on the other hand, not only assist elements of cell mobility but also give the cell form and structure.

## 8.13. CENTRIOLES

The animal cell, which may reproduce or generate duplicates on its own, exemplifies this. This is composed of nine microtubule bundles, and its main job is to help organize the cellular division procedure.

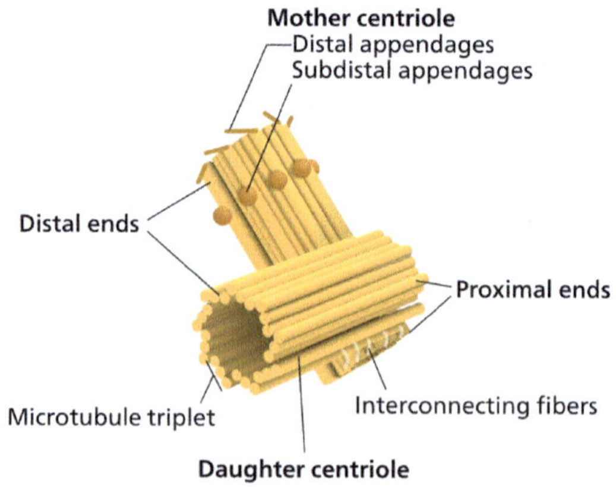


**Figure 8.13.** Centrioles' Structure.

**Source:** https://commons.wikimedia.org/wiki/File:Centriole-en.svg

## 8.13.1. Structure and Functions of Centrioles

This is a tiny structure composed of nine pairs of microtubules arranged in 3 groups, thus the name "triplet microtubules." They are found in structures such as flagella and cilia because they are highly powerful collectively as triplets.

Proteins hold the triplet microtubules together, providing the centriole with its form. They are present in the centrosome and are responsible for the formation and maintenance of microtubules inside the cell. A pericentriolar matrix consisting of a molecule that makes up the microtubules surrounds the triplet microtubules. Tubulin subunits bind with each other to create long-empty cylinders that resemble straw within every microtubule inside the triplet microtubule complex (Lange et al., 1999; Dawe et al., 2007).

The centriole microtubules enable chemicals tied together by a glycoprotein to be transported to every cell site. The glycoprotein linkage serves as a signaling unit that allows certain proteins to be moved.

The microtubules that stretch from the centrioles are anchored by the centrioles, which have the elements required to produce additional tubules.

Mitosis is triggered by the replication of every centriole, which results in the formation of copies of every centriole. The freshly created centrioles split into 2 centrosomes, with every centriole positioned at one angle to the other. The microtubules that connect the centrosomes force the centrioles away, causing them to move to opposite sides of the cell. The microtubules stretch to the cell cytoplasm to find the chromosome after the centrioles are in position. At the centromere, the microtubules attach to the chromosome. The centriole's microtubules are then disassembled, causing the chromosomes to separate (Berns et al., 1977; Basto et al., 2006).

## 8.14. PEROXISOMES

Peroxisomes are important in the oxidation of certain biomolecules. They also play a role in the formation of plasmalogens, which are membrane lipids. Peroxisomes have other activities in plant cells, such as recycling carbon from phosphoglycolate throughout photorespiration. Plants contain specific forms of peroxisomes, such as the glyoxysome, that have been involved in the transformation of fatty acids to carbohydrates.

Peroxisomes have enzymes that oxidize amino acids and fatty acids, which are generally present in the cell. The *peroxisome* gets its name from the $H_2O_2$ produced by these oxidation processes. On the other hand, $H_2O_2$ is potentially hazardous to cells since it may combine with a variety of different chemicals. As a result, peroxisomes include enzymes like catalase, which transform $H_2O_2$ to $H_2O$ and oxygen, thus reducing toxicity. Peroxisomes offer a secure environment for the oxidative metabolism of specific substances in this way (Amano, 1957; Fırat-Karalar & Stearns, 2014).

Human plasmalogens are the major ether lipids (ether lipids differ from conventional lipids in that they include one or more ether links rather than ester connections). The formation of an ether phospholipid precursor is catalyzed by specific enzymes in peroxisomes. Plasmalogen is produced when the precursor molecule undergoes additional manufacturing in the ER. Even though the physiological purpose of plasmalogens is unknown, abnormalities in their production caused by peroxisomal diseases have been linked to significant developmental illnesses such as rhizomelic chondrodysplasia punctata (RCDP) and Zellweger syndrome. Plasmalogen levels in the brain are found to be lower in Alzheimer's patients, which has been associated with cognitive impairment (Chapman et al., 2000; Marshall, 2007).

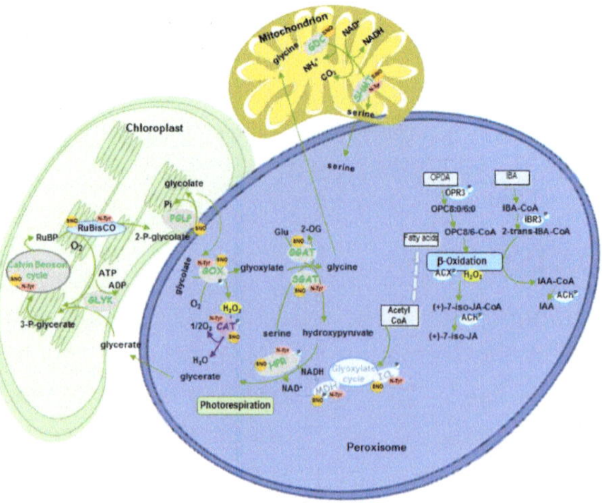


**Figure 8.14.** Peroxisome functions and structure are illustrated in detail.

**Source:** Sandalio LM, Gotor C, Romero LC, Romero-Puertas MC. Multilevel Regulation of Peroxisomal Proteome by Post-Translational Modifications. International Journal of Molecular Sciences. 2019; 20(19):4881. https://doi.org/10.3390/ijms20194881

## 8.14.1. Peroxisomal Disorders

Genes that help make peroxisomes or make the enzymes and transporters that carry the enzymes from the cytoplasm to the peroxisome are the ones that get changed. Peroxisomal disorders are birth defects, and they can be mild to very bad. Zellweger syndrome, newborn adrenoleukodystrophy (NALD), and infantile Refsum disease are all part of the Zellweger spectrum. Zellweger syndrome is when there are no or very few peroxisomes in the body. This is the most serious situation in the Zellweger syndrome, and it is very bad. Zellweger syndrome is caused by mutations that induce Cu, Fe, and substances are known as very long-chain fatty acids to build up in the tissues and blood, including the brain, liver, and kidneys. This build-up can lead to health problems. Zellweger syndrome is a condition that can cause facial deformities and intellectual disabilities in babies. A few may also have hearing and vision problems, and they may have gastrointestinal bleeding or organ failure. Most babies with Zellweger syndrome don't live past the age of one. Signs of NALD and infantile Refsum disease, on the other hand, start to show up in late infancy or early childhood, and patients may live to be young adults. Patients with RCDP can also live into their childhood or their early adulthood in minor cases (van Meer, 1989; Netherton & Wileman, 2011).

Peroxisomes had first been defined in 1960 as a piece of work by Christian René de Duve, who worked on cell fractionation. Using De Duve's method, you can separate organelles depending on how fast they sink to the bottom and how dense they are. Peroxisomes are denser than other organelles. When he was done, he came up with the name *peroxisome*. George Palade and Claude shared the Nobel Prize for Physiology or Medicine in 1974 for their work with De Duve (De Duve et al., 1966; Schrader & Fahimi, 2006).

## 8.15. CILIA AND FLAGELLA

The motile cellular extensions flagella and cilia have been observed in most microbes and animals, although not in high plants. Cilia enable multicellular animal's mobility cells or groups of cells, as well as convey fluids and materials past them. Cilia border the respiratory path in humans, preventing dust, pollution, and potentially hazardous germs from reaching the lungs. Cilia create water currents that convey oxygen and food past the gills of clams and food through the digestive tracts of snails, amongst many other things. Flagella are typically observed on gametes, but they also help coelenterates and sponges breathe and circulate by creating water currents. Flagella and cilia are required for single organism motility in single-celled eukaryotes. Cilia are found on protozoans related to the phylum Ciliophora, whereas flagella are found on protozoans related to the phylum Mastigophora (Roels & Goldfischer, 1979; Akuma et al., 2019 ).

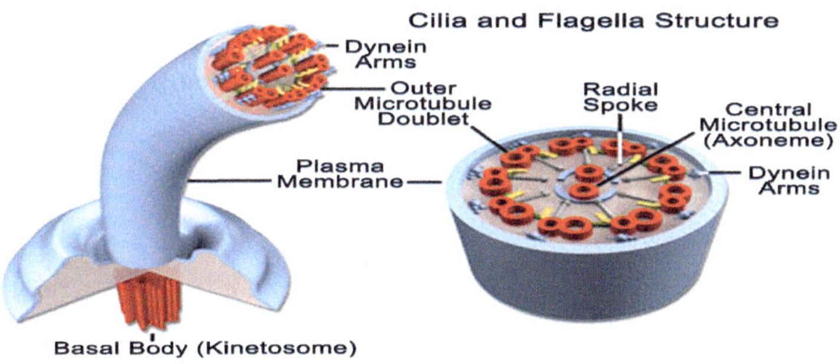


**Figure 8.15.** The structure of Flagella and Cilia is depicted graphically.

**Source:** https://micro.magnet.fsu.edu/cells/ciliaandflagella/ciliaandflagella.html

Flagella and cilia in eukaryotic cells include the motor protein dynein, as well as microtubules, which have been linear polymers of globular proteins known as tubulin. The axoneme is the center of every structure, and it consists of 2 main microtubules encircled by a circle of 9 doublet microtubules. Every doublet microtubule is made up of one whole microtubule and a single partial microtubule, the last of which shares a tubule wall with the other microtubule (shown above ). Dynein molecules are found at regular intervals throughout the perimeter of the axoneme, wherein they bridge the gaps among neighboring microtubule doublets (Zoeller et al., 1988; Honsho et al., 2008).

The complete axoneme complex, which has been connected to the cell at the basal body, is surrounded by a plasma membrane (also called kinetosome). The fundamental external ring shape of the axoneme is preserved in basic bodies, however, each one of the 9 pairs of circular filaments is made up of 3 microtubules instead of a doublet. As a result, the basic body is identical to the centrioles observed in the cell's centrosome, which is positioned closest to the nucleus. Basic bodies have been transformed into centrioles and its flagella are resorbed preceding cellular division in certain species, like the single-celled Chlamydomonas.

Flagella and cilia in eukaryotes have been distinguished by their number and size: cilia are usually smaller and exist in far increased amounts as compared to flagella, which are often isolated. The structures also have slightly distinct forms of motion, albeit movement is created in both situations via dynein stimulation and the resulting twisting of the axoneme. Cilia movement is frequently characterized as whip-like or likened to swimming's breaststroke. Nearby cilia move roughly instantaneously, resulting in wave-like movement patterns in clusters of cilia. Flagella, on the other hand, move in eukaryotes in a uniform, undulatory pattern. The basic motor propels prokaryotic flagella, which have a structure that is fundamentally different from that of the protein flagellin (Silflow & Lefebvre, 2001; Carvalho-Santos et al., 2011).

Disorders in the flagella and cilia of human cells have been linked to several serious medical issues. Issues with the dynein arms which stretch between the microtubules found in the axoneme, for instance, produce Kartagener's syndrome, which is defined via repeated respiratory diseases due to the failure of cilia in the respiratory system to clean out bacteria or other debris. Male sterility is also an outcome of the condition since sperm cells are unable to propel themselves through flagella. Destruction to the respiratory cilia can be gained instead of inherited and is typically connected to smoking cigarettes. Bronchitis, for example, is caused by the construction of mucus and tar in the respiratory system that can't be fully ejected due to cilia disorder caused by smoking (Gilpin et al., 2020).

## 8.16. ENDOSOME

Endosomes seem to be membrane-bound compartments located in the cytoplasm of nearly each animal cell, generated via a complicated series of procedures known as endocytosis. The underlying process of endocytosis is the polar opposite of exocytosis or cell ejection. The plasma membrane of a cellular invaginates (folds inside) to envelop macromolecules or other materials diffusing through the extracellular fluid. The surrounded foreign components are then transported into the cell and released into the cytoplasm in a pouch vesicle after a pinching-off of the membrane (a process known as budding). Vesicles come in a variety of sizes, but greater than one hundred nm in diameter are known as vacuoles (Bettencourt-Dias et al., 2011).

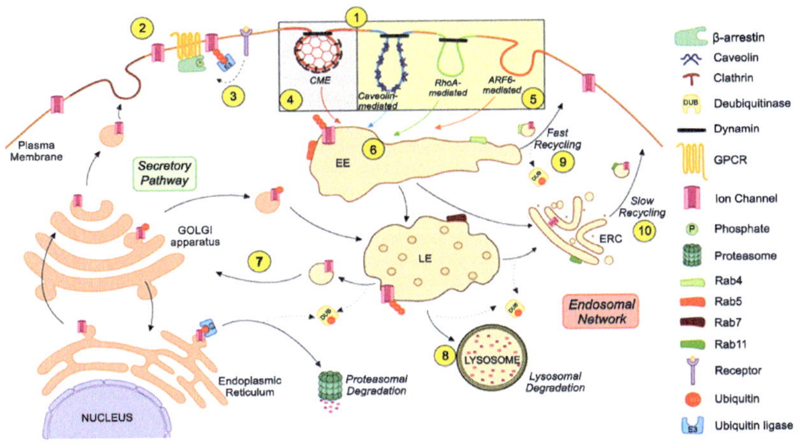


**Figure 8.16.** Endocytosis in Animal Cell.

**Source:** Estadella I, Pedrós-Gámez O, Colomer-Molera M, Bosch M, Sorkin A, Felipe A. Endocytosis: A Turnover *Mechanism Controlling Ion Channel Function.* Cells. *2020; 9(8):1833. https://doi.org/10.3390/cells9081833*

In the diagram above, 3 basic methods of endocytosis that a normal cell exhibits have been depicted. The most directly aimed type of the endocytic mechanism, receptor-mediated endocytosis, is shown on the far left of the diagram. Active cells may take in large amounts of certain chemicals (ligands) which link to receptor sites stretching from the cytoplasmic membrane into the extracellular fluid near the cell via receptor-mediated endocytosis. Coating pits in the membrane, which have been lined on its cytoplasmic side with a bristle-like set of coat proteins, are frequently associated with these receptor sites. The coat proteins have been hypothesized to aid in the formation of vesicles by expanding the pit. Note that vesicles formed by receptor-mediated endocytosis can import other molecules in addition to ligands, as illustrated in the diagram, albeit ligands are normally taken into the cell in larger concentrations (Fritz-Laylin, 2020).

Pinocytosis, which has been seen in the centre area of Figure 1, is a somewhat selective method of endocytosis. A cell may consume liquid droplets from the extracellular fluid through pinocytosis. All solutes contained in extracellular droplets can get enclosed in the vesicles created as a result of such a procedure, with those located in the highest concentration in the extracellular fluid also becoming the most abundant in the membrane sacs. Various endocytic mechanisms create vesicles that are shorter than pinocytic vesicles (Witman, 1990).

The fourth kind of endocytosis, phagocytosis (shown in the figure below), is arguably a very well way for a cell to receive things from the outside world. Children see amoebas underneath the microscope in several school science laboratories, watching the unicellular creatures consume via extending out pseudopodia and surrounding all food particles that cross its pathway. Phagocytosis is the process of engulfing particles and packing them into vesicles, which are generally big enough to be called vacuoles. Despite its association with amoebas, phagocytosis is used by a wide range of species. In most multicellular creatures, phagocytic cells serve primarily as a tool of biological defense instead of as a source of nutrition. Protozoa, dead cells, bacteria, and other items are phagocytosed by leukocytes in the mammalian body to assist avoid infections and other disorders (Fawcett, 1961).

Multiple tiny vesicles formed by endocytosis can merge to create a single individual once released into the cytoplasm. In most cases, this endosome performs one of two tasks. Endosomes typically transfer their components in a sequence of stages to a lysosome, that digests the components. Endosomes, on the other hand, are employed by the cell to transfer diverse chemicals across various parts of the exterior cell membrane in other cases. Since epithelial cells, including those which make up the external layer of the skin, have polarization ability, such function is very significant (One side of the cell is not the same as the other). A fluorescence digital picture of a single African green monkey kidney fibroblast cell (CV-1 line) injected with a fluorescent protein linked to an endosome directing amino acid pattern is shown in the Figure. In the diagram, the plasma membrane, the nucleus, and endosome constituents are labeled (Lemmon & Traub, 2000).

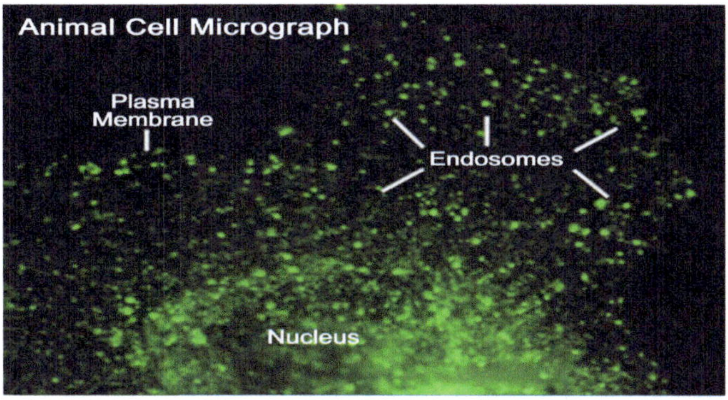


**Figure 8.17.** The plasma membrane, the nucleus, and the endosomes are all visible in this micrograph of an animal cell.

**Source:** https://micro.magnet.fsu.edu/cells/endosomes/endosomes.html

An endosome that is on its approach to transferring its components to a lysosome usually undergoes various alterations. The specialized vesicle includes a single segment in its earliest phase, once it is known as an initial endosome. Chemical variations in the vesicle, on the other hand, eventually occur, and the membrane enveloping the endosome bends in on itself in a similar manner to plasma membrane invagination. The membrane, on the other hand, is not squeezed off in this circumstance. As a result, a multivesicular endosome, a structure having several sections, is created. The multivesicular endosome is an intermediary architecture wherein the vesicle undergoes further chemical changes, such as a considerable drop in pH level, as it develops into a late endosome.

Even though late endosomes can decompose a wide range of proteins and lipids, a lysosome is required to completely digest all of the contents they comprise. The components of late endosomes are commonly transported to lysosomes via the fusion of their membranes. Late endosomes may grow into lysosomes by further structural and chemical alterations in certain conditions, and fusion is not required for digestion to be finished (Dupre et al., 2004).

## 8.17. VACUOLES

A vacuole is a cellular organelle that stores different solutions or materials. This comprises both solutions that are being generated and are stored or expelled, as well as those that are phagocytized, or absorbed by the cell. A vacuole is essentially a compartment enclosed by a membrane that prevents the cytosol from coming into contact with the substances within. Vacuoles have been enclosed by semi-permeable membranes that allow only specific molecules to pass through (Montagnac et al., 2003).

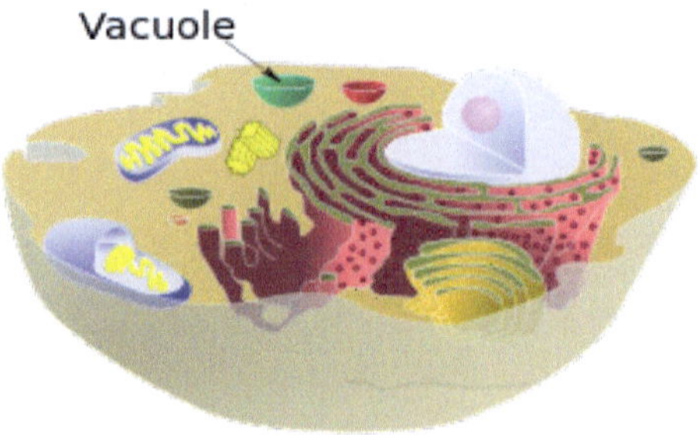


**Figure 8.18.** Animal Cell is depicted graphically.

**Source:** https://www.britannica.com/science/vacuole

### 8.17.1. Structure of Vacuole

A vacuole is a term that contains a wide range of membrane-bound sacs. Phospholipids make up the membranes; however, every creature can employ bit distinct phospholipids. Proteins are implanted in membranes and may either transport chemicals throughout the membrane or provide its structure. Distinct protein pairings enable various vacuoles to manage and retain various materials.

Various genetics induce various proteins to attach to the vacuole membrane, allowing distinct chemicals to pass through and giving the vacuoles various characteristics. Vacuoles have developed to serve as water storage organelles in most plant cells, and they also serve a range of other roles. Animals do not depend upon this water storage for rigidity of form, instead of using its vacuoles for storage of goods, endocytosis, and exocytosis (Finlay & Falkow, 1988).

## 8.17.2. Functions of Vacuole

### *8.17.2.1. Water Storage*

A big vacuole takes up the bulk of the cell space in plants. The tonoplast, a form of the cytoplasmic membrane which may expand and fill itself with a fluid referred to as cell sap, surrounds such vacuole. The vacuole also absorbs protons from the cytoplasm, resulting in an acidic environment within the cell. The chemical gradient formed by the vacuole may then be used to transfer molecules into and outside of the vacuole, a process known as proton motive force. It involves water and another molecular movement (Baluška, 2006). The vacuole within a plant cell is seen in the image below.

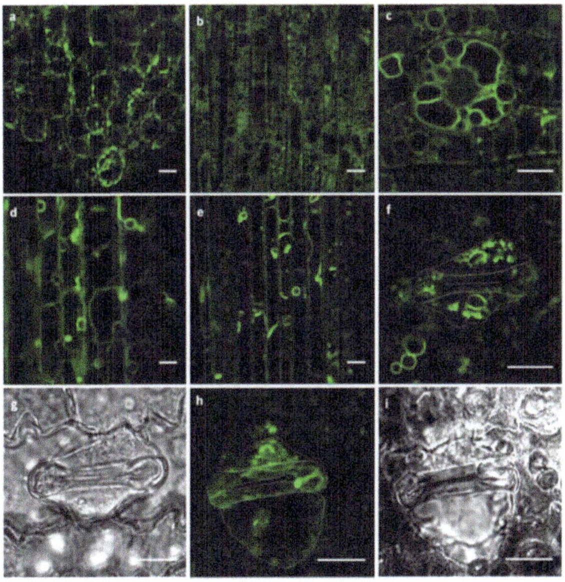

**Figure 8.19.** This diagram depicts the Vacuole's water storage.

*Source: Cao Y, Cai W, Chen X, Chen M, Chu J, Liang W, Persson S, Liu Z, Zhang D. Bright Fluorescent Vacuolar Marker Lines Allow Vacuolar Tracing Across Multiple Tissues and Stress Conditions in Rice.* International Journal of Molecular Sciences. *2020; 21(12):4203. https://doi.org/10.3390/ijms21124203*

(**a**,**b**) Coleoptile epidermal cells are enriched with small bubble-like vacuoles. Coleoptiles, five days post-germination, were used to take images from the tip (**a**) and middle (**b**) regions. (**c**) The stomata in the coleoptile was packed with fragmented vacuoles. (**d**,**e**) Central vacuoles in true leaves. The big central vac-uoles occupied most of the leaf pavement cells (**d**) and developing veins (**e**). (**f**)

Vacuoles in closed stomata of true leaf. (**g**) The corresponding bright field channel of (**f**). (**h**) Vacuoles in opened stomata of true leaf. (**i**) The corresponding bright field channel of (**h**). Yellow and blue lines indicate the cell boundaries of subsidiary cells and guard cells, respectively. White lines indicate the pore region of the stomata. Scale bars: 10 μm (Cao et al, 2020).

### *8.17.2.2. Turgor Pressure*

Plants utilize their vacuoles for a 2nd purpose that is critical to their survival. Whenever submerged in water, the vacuole becomes pressured and exerts pressure on the cell walls. Even though the turgor pressure in every cell is modest, it permits the cells to create a shape and withstand rain, wind, and even hail. While woody plants produce more fibers and proteins to assist them to stand taller, several non-woody plants may grow to be several feet tall just by turgor pressure.

Although it is an effective technique for plants to organize themselves, if the pH balance is incorrect or there isn't enough water, the plant may drop. If your plants are drooping, ensure they're getting enough water. In a couple of hours, they may go from wilting to turgid with a fast sip. Assess the pH of the soil when that does not work. The roots cannot gain water or nutrients to deposit in the vacuoles of the cells if the circumstances aren't appropriate, and the plant wilts. If you change the pH, the plant would tighten up immediately (Kobayashi et al., 1998).

### *8.17.2.3. Endocytosis and Exocytosis*

The term "vacuole" refers to a structure that is employed when a considerable volume of material is taken in by endocytosis or expelled via exocytosis. Animals and plant cells alike take in chemicals that should be stored apart from the cytoplasm to function properly. This may be because the compounds are reactive, wherein scenario they would create undesirable reactions. It's possible that the compounds are being avoided because they will interfere with biological activities. After all, they are both rather substantial. Lysosomes are vesicles that are responsible for ingesting chemicals that will be digested. Occasionally, these lysosomes may join together to create a big digestive vesicle that may metabolize nutrients in an acidic medium before transferring them to the cytosol or other organelles for usage. This is known as fusion. Endocytosis is the term used to describe such a procedure, which differs across various kinds of cells. In the other direction, numerous cells perform the role of secretory cells, which are responsible for the production and excretion of enormous quantities of various chemicals. The chemicals are created in the ER and transported to the Golgi apparatus, where they are changed and labeled in preparation for distribution to the rest of the body. After that, the chemicals may be placed into vesicles. The vesicles can migrate through the cytoplasm and unite into a bigger vacuole before being ejected. Exocytosis is the term used to describe this process. The shape of the vacuoles that transport various chemicals between cells, and even between cells whenever they perform distinct activities, vary from one cell to the next. A single animal cell may have a large number of vacuoles, each of which performs a different purpose. An illustration of an animal cell, as well as its vacuoles, may be found below; the small unlabeled spheres represent vesicles, and the larger labeled spheres represent vacuoles. Once they have fused, they will be termed vacuoles as well (Gruenberg, 2009).

### 8.17.2.4. Storage Functions

Vacuoles are capable of storing a wide range of various sorts of chemicals. Fat cells, for example, may store enormous quantities of lipids in specific vacuoles within their walls. This enables single cells to store a significant quantity of fat, which animals may utilize if sources are few or unavailable. Because of the extensibility of the vacuole, an organism may increase or lose weight without creating or destroying an excessive number of cells. Other times, the vacuoles of organisms have been employed to construct complete ecosystems wherein symbiotic species can coexist peacefully. Coral polyps frequently consume algae through the process of endocytosis, yet the algae are permitted to persist in the coral's vacuoles. This enables the coral to absorb the nutrients and oxygen released by the algae and use them to grow.

## 8.18. MICROVILLI

Microvilli are small, narrow protrusions that can be observed on the free surfaces of certain differentiated cells; they are more common in epithelial cells than in other differentiated cells. They are filiform structures with lengths ranging from one to two micrometers and thicknesses ranging from one hundred to two hundred nanometers. On the inside of every microvilli, hundreds of actin filaments are arranged parallel to the long axis of the cell. Microvilli are often thick and densely packed, similar to the way a brush edge is packed. Microvilli have been spatially structured as hexagons in the top aspect of the cell.

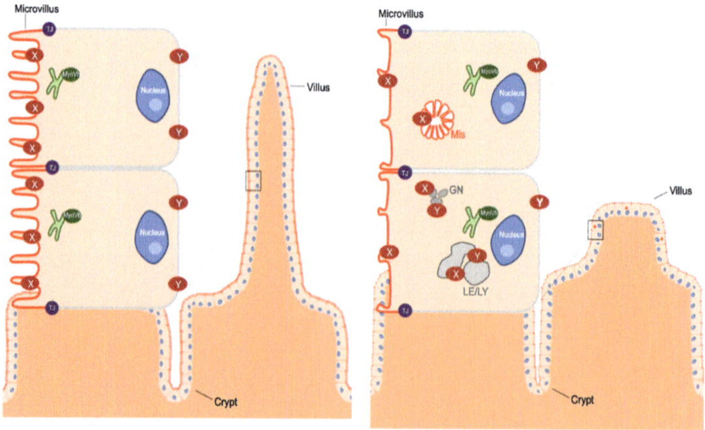


Normal MVID

**Figure 8.20.** Microvilli in animal cells are depicted graphically.

**Source:** Leng C, Rings EHHM, de Wildt SN, van IJzendoorn SCD. Pharmacological and Parenteral Nutrition-Based Interventions in Microvillus Inclusion Disease. Journal of Clinical Medicine. 2021; 10(1):22. https://doi.org/10.3390/jcm10010022

## 8.18.1. Structure and Functions of Microvilli

These are protrusions on the surface of the actin filaments that are generated by accessory proteins. Microvilli are formed on the surface of the cellular membrane as a result of the association of extra proteins.

They work by increasing the surface area of the small intestine, which allows for better uptake of water and metabolized food. Certain microvilli can be discovered in the ear, which is responsible for the detection of sound and the transmission of sound waves to the brain through an electric signal. They also aid in the attachment of the sperm to the egg, allowing for easier fertilization. Additionally, anchors are found in white blood cells, enabling them to circulate freely through the circulatory system to connect to potential infections as they go (Marcote et al., 2000).

## REFERENCES

1. Akhmanova, A., & Kapitein, L. C. (2022). Mechanisms of microtubule organization in differentiated animal cells. *Nature Reviews Molecular Cell Biology*, 1-18.
2. Akuma, P., Okagu, O. D., & Udenigwe, C. C. (2019). Naturally occurring exosome vesicles as potential delivery vehicle for bioactive compounds. *Frontiers in Sustainable Food Systems, 3*, 23.
3. Amano, S. (1957). The Structure of the Centrioles and Spindle Body as Observed Under the Electron and Phase Contrast Microscope SA New Extension-fiber Theory Concerning Mitotic Mechanism in Animal Cells. *Cytologia, 22*(2), 193-212.
4. Ambros, V. (2004). The functions of animal microRNAs. *Nature, 431*(7006), 350-355.
5. Baluška, F., Menzel, D., & Barlow, P. W. (2006). Cytokinesis in plant and animal cells: endosomes 'shut the door'. *Developmental biology, 294*(1), 1-10.
6. Barak, L. S., Yocum, R. R., Nothnagel, E. A., & Webb, W. W. (1980). Fluorescence staining of the actin cytoskeleton in living cells with 7-nitrobenz-2-oxa-1, 3-diazole-phallacidin. *Proceedings of the National Academy of Sciences, 77*(2), 980-984.
7. Barritt, G. J. (1999). Receptor-activated Ca2+ inflow in animal cells: a variety of pathways tailored to meet different intracellular Ca2+ signalling requirements. *Biochemical Journal, 337*(2), 153-169.
8. Basto, R., Lau, J., Vinogradova, T., Gardiol, A., Woods, C. G., Khodjakov, A., & Raff, J. W. (2006). Flies without centrioles. *Cell, 125*(7), 1375-1386.
9. Bernardino de la Serna, J., Schütz, G. J., Eggeling, C., & Cebecauer, M. (2016). There is no simple model of the plasma membrane organization. *Frontiers in cell and developmental biology, 4*, 106.
10. Berns, M. W., Rattner, J. B., Brenner, S., & Meredith, S. (1977). The role of the centriolar region in animal cell mitosis. A laser microbeam study. *The Journal of cell biology, 72*(2), 351-367.
11. Bettencourt-Dias, M., Hildebrandt, F., Pellman, D., Woods, G., & Godinho, S. A. (2011). Centrosomes and cilia in human disease. *Trends in Genetics, 27*(8), 307-315.
12. Bhattacharyya, D., & Glick, B. S. (2007). Two mammalian Sec16 homologues have nonredundant functions in endoplasmic reticulum

(ER) export and transitional ER organization. *Molecular biology of the cell*, *18*(3), 839-849.

13. Boon, K., Caron, H. N., Van Asperen, R., Valentijn, L., Hermus, M. C., Van Sluis, P., ... & Versteeg, R. (2001). N-myc enhances the expression of a large set of genes functioning in ribosome biogenesis and protein synthesis. *The EMBO journal*, *20*(6), 1383-1393.
14. Borst, P. (1977). Structure and function of mitochondrial DNA. *Trends in Biochemical Sciences*, *2*(2), 31-34.
15. Boyko, V., Ferralli, J., Ashby, J., Schellenbaum, P., & Heinlein, M. (2000). Function of microtubules in intercellular transport of plant virus RNA. *Nature Cell Biology*, *2*(11), 826-832.
16. Bray, D., & White, J. G. (1988). Cortical flow in animal cells. *Science*, *239*(4842), 883-888.
17. Butler, M. (2005). Animal cell cultures: recent achievements and perspectives in the production of biopharmaceuticals. *Applied microbiology and biotechnology*, *68*(3), 283-291.
18. Caldarola, S., De Stefano, M. C., Amaldi, F., & Loreni, F. (2009). Synthesis and function of ribosomal proteins–fading models and new perspectives. *The FEBS Journal*, *276*(12), 3199-3210.
19. Carraway, K. L., & Carraway, C. A. C. (1989). Membrane-cytoskeleton interactions in animal cells. *Biochimica et Biophysica Acta (BBA)-Reviews on Biomembranes*, *988*(2), 147-171.
20. Carvalho-Santos, Z., Azimzadeh, J., Pereira-Leal, J. B., & Bettencourt-Dias, M. (2011). Evolution: Tracing the origins of centrioles, cilia, and flagella. *The Journal of cell biology*, *194*(2), 165.
21. Chapman, M. J., Dolan, M. F., & Margulis, L. (2000). Centrioles and kinetosomes: form, function, and evolution. *The Quarterly review of biology*, *75*(4), 409-429.
22. Clegg, J. S. (1984). Properties and metabolism of the aqueous cytoplasm and its boundaries. *American Journal of Physiology-Regulatory, Integrative and Comparative Physiology*, *246*(2), R133-R151.
23. Clément, P. (2007). Introducing the cell concept with both animal and plant cells: A historical and didactic approach. *Science & Education*, *16*(3), 423-440.
24. Conduit, P. T., Wainman, A., & Raff, J. W. (2015). Centrosome function and assembly in animal cells. *Nature reviews Molecular cell biology*, *16*(10), 611-624.

25. Conduit, P. T., Wainman, A., & Raff, J. W. (2015). Centrosome function and assembly in animal cells. *Nature reviews Molecular cell biology*, *16*(10), 611-624.

26. Dawe, H. R., Farr, H., & Gull, K. (2007). Centriole/basal body morphogenesis and migration during ciliogenesis in animal cells. *Journal of cell science*, *120*(1), 7-15.

27. De Duve, C. A. B. P., & Baudhuin, P. (1966). Peroxisomes (microbodies and related particles). *Physiological Reviews*, *46*(2), 323-357.

28. de la Cruz, J., Karbstein, K., & Woolford Jr, J. L. (2015). Functions of ribosomal proteins in assembly of eukaryotic ribosomes in vivo. *Annual review of biochemistry*, *84*, 93-129.

29. Di Leva, F., Domi, T., Fedrizzi, L., Lim, D., & Carafoli, E. (2008). The plasma membrane Ca2+ ATPase of animal cells: structure, function and regulation. *Archives of biochemistry and biophysics*, *476*(1), 65-74.

30. Dong, Y., Yoshitomi, T., Hu, J. F., & Cui, J. (2017). Long noncoding RNAs coordinate functions between mitochondria and the nucleus. *Epigenetics & chromatin*, *10*(1), 1-11.

31. Driouich, A., Faye, L., & Staehelin, A. (1993). The plant Golgi apparatus: a factory for complex polysaccharides and glycoproteins. *Trends in biochemical sciences*, *18*(6), 210-214.

32. Dupre, S., Urban-Grimal, D., & Haguenauer-Tsapis, R. (2004). Ubiquitin and endocytic internalization in yeast and animal cells. *Biochimica et Biophysica Acta (BBA)-Molecular Cell Research*, *1695*(1-3), 89-111.

33. Eizirik, D. L., Cardozo, A. K., & Cnop, M. (2008). The role for endoplasmic reticulum stress in diabetes mellitus. *Endocrine Reviews*, *29*(1), 42-61.

34. Fawcett, D. (1961). Cilia and flagella. In *The cell* (pp. 217-297). Academic Press.

35. Fededa, J. P., & Gerlich, D. W. (2012). Molecular control of animal cell cytokinesis. *Nature cell biology*, *14*(5), 440-447.

36. Finlay, B. B., & Falkow, S. (1988). Comparison of the invasion strategies used by *Salmonella cholerae-suis*, *Shigella flexneri* and *Yersinia enterocolitica* to enter cultured animal cells: endosome acidification is not required for bacterial invasion or intracellular replication. *Biochimie*, *70*(8), 1089-1099.

37. Fırat-Karalar, E. N., & Stearns, T. (2014). The centriole duplication cycle. *Philosophical Transactions of the Royal Society B: Biological*

*Sciences, 369*(1650), 20130460.

38. Franke, W. W., Kartenbeck, J., Krien, S., Vander Woude, W. J., Scheer, U., & Morre, D. J. (1972). Inter-and intracisternal elements of the Golgi apparatus. *Zeitschrift für Zellforschung und Mikroskopische Anatomie, 132*(3), 365-380.
39. Friedman, J. R., DiBenedetto, J. R., West, M., Rowland, A. A., & Voeltz, G. K. (2013). Endoplasmic reticulum–endosome contact increases as endosomes traffic and mature. *Molecular biology of the cell, 24*(7), 1030-1040.
40. Fritz-Laylin, L. K. (2020). The evolution of animal cell motility. *Current Biology, 30*(10), R477-R482.
41. Fu, S., Watkins, S. M., & Hotamisligil, G. S. (2012). The role of endoplasmic reticulum in hepatic lipid homeostasis and stress signaling. *Cell metabolism, 15*(5), 623-634.
42. Gilpin, W., Bull, M. S., & Prakash, M. (2020). The multiscale physics of cilia and flagella. *Nature Reviews Physics, 2*(2), 74-88.
43. Goldberg, A. L. (1993). The mechanism and functions of ATP-dependent proteases in bacterial and animal cells. *EJB Reviews*, 1-15.
44. Gonzalez, Y., Saito, A., & Sazer, S. (2012). Fission yeast Lem2 and Man1 perform fundamental functions of the animal cell nuclear lamina. *Nucleus, 3*(1), 60-76.
45. Gościńska, K., & Topf, U. (2020). The evolving role of ribosomes in the regulation of protein synthesis. *Acta Biochimica Polonica, 67*(4), 465-473.
46. Gruenberg, J. (2009). Viruses and endosome membrane dynamics. *Current opinion in cell biology, 21*(4), 582-588.
47. Hall, A. (2009). The cytoskeleton and cancer. *Cancer and Metastasis Reviews, 28*(1), 5-14.
48. Harder, T., Scheiffele, P., Verkade, P., & Simons, K. (1998). Lipid domain structure of the plasma membrane revealed by patching of membrane components. *The Journal of cell biology, 141*(4), 929-942.
49. Hartley, M. R., Chaddock, J. A., & Bonness, M. S. (1996). The structure and function of ribosome-inactivating proteins. *Trends in Plant Science, 1*(8), 252.
50. Hibbard, H. (1945). Current status of our knowledge of the Golgi apparatus in the animal cell. *The Quarterly Review of Biology, 20*(1), 1-19.

51. Honsho, M., Yagita, Y., Kinoshita, N., & Fujiki, Y. (2008). Isolation and characterization of mutant animal cell line defective in alkyl-dihydroxyacetonephosphate synthase: localization and transport of plasmalogens to post-Golgi compartments. *Biochimica et Biophysica Acta (BBA)-Molecular Cell Research, 1783*(10), 1857-1865.
52. Ierushalmi, N., & Keren, K. (2021). Cytoskeletal symmetry breaking in animal cells. *Current Opinion in Cell Biology, 72*, 91-99.
53. Ingólfsson, H. I., Melo, M. N., Van Eerden, F. J., Arnarez, C., Lopez, C. A., Wassenaar, T. A., ... & Marrink, S. J. (2014). Lipid organization of the plasma membrane. *Journal of the american chemical society, 136*(41), 14554-14559.
54. Ishikawa, T., Li, Z. S., Lu, Y. P., & Rea, P. A. (1997). The GS-X pump in plant, yeast, and animal cells: structure, function, and gene expression. *Bioscience reports, 17*(2), 189-207.
55. Jacobs, C. W., Adams, A. E., Szaniszlo, P. J., & Pringle, J. R. (1988). Functions of microtubules in the Saccharomyces cerevisiae cell cycle. *The Journal of cell biology, 107*(4), 1409-1426.
56. Jimenez, A., Chen, D., & Alto, N. M. (2016). How bacteria subvert animal cell structure and function. *Annual review of cell and developmental biology, 32*, 373-397.
57. Joazeiro, C. A. (2019). Mechanisms and functions of ribosome-associated protein quality control. *Nature Reviews Molecular Cell Biology, 20*(6), 368-383.
58. Karsenti, E., & Maro, B. (1986). Centrosomes and the spatial distribution of microtubules in animal cells. *Trends in Biochemical Sciences, 11*(11), 460-463.
59. Kirillova, A., Smitz, J. E., Sukhikh, G. T., & Mazunin, I. (2021). The role of mitochondria in oocyte maturation. *Cells, 10*(9), 2484.
60. Kite, G. L. (1913). The Relative Permeability Of The Surface And Interior Portions Of The Cytoplasm Of Animal And Plant Cells: (A Preliminary Paper.). *The Biological Bulletin, 25*(1), 1-7.
61. Kite, G. L. (1915). Studies on the permeability of the internal cytoplasm of animal and plant cells. *American Journal of Physiology-Legacy Content, 37*(2), 282-299.
62. Kobayashi, T., Stang, E., Fang, K. S., de Moerloose, P., Parton, R. G., & Gruenberg, J. (1998). A lipid associated with the

antiphospholipid syndrome regulates endosome structure and function. *Nature, 392*(6672), 193-197.

63. Koch, J., & Götz, D. (1972). Transport of nuclear DNA into the cytoplasm in animal cells The complexity of the transported DNA. *FEBS letters, 27*(1), 9-12.
64. Lange, B. M. H., Faragher, A. J., March, P., & Gull, K. (1999). Centriole duplication and maturation in animal cells. *Current topics in developmental biology, 49*, 235-249.
65. Lange, B. M., & Gull, K. (1996). Structure and function of the centriole in animal cells: progress and questions. *Trends in cell biology, 6*(9), 348-352.
66. Lee, H. C., & Wei, Y. H. (1997). Role of mitochondria in human aging. *Journal of biomedical science, 4*(6), 319-326.
67. Lemmon, S. K., & Traub, L. M. (2000). Sorting in the endosomal system in yeast and animal cells. *Current opinion in cell biology, 12*(4), 457-466.
68. Lenaz, G., Baracca, A., Fato, R., Genova, M. L., & Solaini, G. (2006). New insights into structure and function of mitochondria and their role in aging and disease. *Antioxidants & redox signaling, 8*(3-4), 417-437.
69. Li, X. D., Wu, J., Gao, D., Wang, H., Sun, L., & Chen, Z. J. (2013). Pivotal roles of cGAS-cGAMP signaling in antiviral defense and immune adjuvant effects. *Science, 341*(6152), 1390-1394.
70. Luna, E. J., & Hitt, A. L. (1992). Cytoskeleton—plasma membrane interactions. *Science, 258*(5084), 955-964.
71. Lynes, E. M., & Simmen, T. (2011). Urban planning of the endoplasmic reticulum (ER): how diverse mechanisms segregate the many functions of the ER. *Biochimica Et Biophysica Acta (BBA)-Molecular Cell Research, 1813*(10), 1893-1905.
72. Magnuson, B., Ekim, B., & Fingar, D. C. (2012). Regulation and function of ribosomal protein S6 kinase (S6K) within mTOR signalling networks. *Biochemical Journal, 441*(1), 1-21.
73. Mao, J., Hu, Y., Ruan, L., Ji, Y., & Lou, Z. (2019). Role of endoplasmic reticulum stress in depression. *Molecular Medicine Reports, 20*(6), 4774-4780.
74. Marcote, M. J., Gu, F., Gruenberg, J., & Aniento, F. (2000). Membrane transport in the endocytic pathway: animal versus plant cells. *Protoplasma, 210*(3), 123-132.

75. Marshall, W. F. (2007). What is the function of centrioles?. *Journal of cellular biochemistry*, *100*(4), 916-922.

76. Maxfield, F. R. (2002). Plasma membrane microdomains. *Current opinion in cell biology*, *14*(4), 483-487.

77. Mayran, N., Parton, R. G., & Gruenberg, J. (2003). Annexin II regulates multivesicular endosome biogenesis in the degradation pathway of animal cells. *The EMBO journal*, *22*(13), 3242-3253.

78. Mège, R. M., Gavard, J., & Lambert, M. (2006). Regulation of cell–cell junctions by the cytoskeleton. *Current opinion in cell biology*, *18*(5), 541-548.

79. Minibayeva, F., Dmitrieva, S., Ponomareva, A., & Ryabovol, V. (2012). Oxidative stress-induced autophagy in plants: the role of mitochondria. *Plant Physiology and Biochemistry*, *59*, 11-19.

80. Minogue, J., Gail Jones, M., Broadwell, B., & Oppewall, T. (2006). The impact of haptic augmentation on middle school students' conceptions of the animal cell. *Virtual Reality*, *10*(3), 293-305.

81. Mitchison, T. J., Charras, G. T., & Mahadevan, L. (2008, June). Implications of a poroelastic cytoplasm for the dynamics of animal cell shape. In *Seminars in cell & developmental biology* (Vol. 19, No. 3, pp. 215-223). Academic Press.

82. Moeendarbary, E., Valon, L., Fritzsche, M., Harris, A. R., Moulding, D. A., Thrasher, A. J., ... & Charras, G. T. (2013). The cytoplasm of living cells behaves as a poroelastic material. *Nature materials*, *12*(3), 253-261.

83. Montagnac, G., Echard, A., & Chavrier, P. (2008). Endocytic traffic in animal cell cytokinesis. *Current opinion in cell biology*, *20*(4), 454-461.

84. Murshid, A., & Presley, J. F. (2004). ER-to-Golgi transport and cytoskeletal interactions in animal cells. *Cellular and Molecular Life Sciences CMLS*, *61*(2), 133-145.

85. Nakatsu, F., Baskin, J. M., Chung, J., Tanner, L. B., Shui, G., Lee, S. Y., ... & De Camilli, P. (2012). PtdIns4P synthesis by PI4KIIIα at the plasma membrane and its impact on plasma membrane identity. *Journal of Cell Biology*, *199*(6), 1003-1016.

86. Netherton, C. L., & Wileman, T. (2011). Virus factories, double membrane vesicles and viroplasm generated in animal cells. *Current opinion in virology*, *1*(5), 381-387.

87. Nicolson, G. L. (1974). The interactions of lectins with animal cell surfaces. *International review of cytology, 39*, 89-190.
88. Nilsen, T. W. (2007). Mechanisms of microRNA-mediated gene regulation in animal cells. *TRENDS in Genetics, 23*(5), 243-249.
89. Oakes, S. A., & Papa, F. R. (2015). The role of endoplasmic reticulum stress in human pathology. *Annual Review of Pathology: Mechanisms of Disease, 10*, 173-194.
90. Pagano, R. E. (1988). A fluorescent derivative of ceramide: physical properties and use in studying the Golgi apparatus of animal cells. In *Methods in cell biology* (Vol. 29, pp. 75-85). Academic Press.
91. Pagano, R. E., Martin, O. C., Kang, H. C., & Haugland, R. P. (1991). A novel fluorescent ceramide analogue for studying membrane traffic in animal cells: accumulation at the Golgi apparatus results in altered spectral properties of the sphingolipid precursor. *The Journal of cell biology, 113*(6), 1267-1279.
92. Paoletti, A., Moudjou, M., Paintrand, M., Salisbury, J. L., & Bornens, M. (1996). Most of centrin in animal cells is not centrosome-associated and centrosomal centrin is confined to the distal lumen of centrioles. *Journal of cell science, 109*(13), 3089-3102.
93. Picone, R., Ren, X., Ivanovitch, K. D., Clarke, J. D., McKendry, R. A., & Baum, B. (2010). A polarised population of dynamic microtubules mediates homeostatic length control in animal cells. *PLoS biology, 8*(11), e1000542.
94. Prenner, E. J., Kiricsi, M., Jelokhani-Niaraki, M., Lewis, R. N., Hodges, R. S., & McElhaney, R. N. (2005). Structure-activity relationships of diastereomeric lysine ring size analogs of the antimicrobial peptide gramicidin S: mechanism of action and discrimination between bacterial and animal cell membranes. *Journal of Biological Chemistry, 280*(3), 2002-2011.
95. Ravikumar, B., Moreau, K., Jahreiss, L., Puri, C., & Rubinsztein, D. C. (2010). Plasma membrane contributes to the formation of pre-autophagosomal structures. *Nature cell biology, 12*(8), 747-757.
96. Reddy, P. H. (2009). Role of mitochondria in neurodegenerative diseases: mitochondria as a therapeutic target in Alzheimer's disease. *CNS spectrums, 14*(S7), 8-13.
97. Roels, F., & Goldfischer, S. (1979). Cytochemistry of human catalase. The demonstration of hepatic and renal peroxisomes by a high temperature

procedure. *Journal of Histochemistry & Cytochemistry*, *27*(11), 1471-1477.

98. Rose, J. K., Buonocore, L., & Whitt, M. A. (1991). A new cationic liposome reagent mediating nearly quantitative transfection of animal cells. *Biotechniques*, *10*(4), 520-525.
99. Ross, A. P., & Zarbalis, K. (2014). The emerging roles of ribosome biogenesis in craniofacial development. *Frontiers in physiology*, *5*, 26.
100. Sakai, Y., Yoshida, S., Yoshiura, Y., Mori, R., Tamura, T., Yahiro, K., ... & Nakazawa, K. (2010). Effect of microwell chip structure on cell microsphere production of various animal cells. *Journal of bioscience and bioengineering*, *110*(2), 223-229.
101. Sasabe, M., & Machida, Y. (2012). Regulation of organization and function of microtubules by the mitogen-activated protein kinase cascade during plant cytokinesis. *Cytoskeleton*, *69*(11), 913-918.
102. Satiat-Jeunemaitre, B., & Hawes, C. (1992). Reversible dissociation of the plant Golgi apparatus by brefeldin A. *Biology of the Cell*, *74*, 325-328.
103. Schrader, M., & Fahimi, H. D. (2006). Peroxisomes and oxidative stress. *Biochimica et Biophysica Acta (BBA)-Molecular Cell Research*, *1763*(12), 1755-1766.
104. Seemann, J., Jokitalo, E., Pypaert, M., & Warren, G. (2000). Matrix proteins can generate the higher order architecture of the Golgi apparatus. *Nature*, *407*(6807), 1022-1026.
105. Seemann, J., Pypaert, M., Taguchi, T., Malsam, J., & Warren, G. (2002). Partitioning of the matrix fraction of the Golgi apparatus during mitosis in animal cells. *Science*, *295*(5556), 848-851.
106. Seo, A. Y., Joseph, A. M., Dutta, D., Hwang, J. C., Aris, J. P., & Leeuwenburgh, C. (2010). New insights into the role of mitochondria in aging: mitochondrial dynamics and more. *Journal of cell science*, *123*(15), 2533-2542.
107. Serrano, R. (1989). Structure and function of plasma membrane ATPase. *Annual review of plant biology*, *40*(1), 61-94.
108. Silflow, C. D., & Lefebvre, P. A. (2001). Assembly and motility of eukaryotic cilia and flagella. Lessons from *Chlamydomonas reinhardtii*. *Plant physiology*, *127*(4), 1500-1507.
109. Smirnov, A., Entelis, N., Martin, R. P., & Tarassov, I. (2011). Biological significance of 5S rRNA import into human mitochondria: role of

ribosomal protein MRP-L18. *Genes & development*, *25*(12), 1289-1305.

110. Steinman, R. M., Mellman, I. S., Muller, W. A., & Cohn, Z. A. (1983). Endocytosis and the recycling of plasma membrane. *The Journal of cell biology*, *96*(1), 1-27.
111. Stephens, R. E., & Edds, K. T. (1976). Microtubules: structure, chemistry, and function. *Physiological Reviews*, *56*(4), 709-777.
112. Stewart, M. J., & Thomas, G. (1994). Mitogenesis and protein synthesis: a role for ribosomal protein S6 phosphorylation?. *Bioessays*, *16*(11), 809-815.
113. Tabas, I. (2010). The role of endoplasmic reticulum stress in the progression of atherosclerosis. *Circulation research*, *107*(7), 839-850.
114. Threadgold, L. T. (2017). *The ultrastructure of the animal cell: international series in pure and applied biology* (Vol. 55, pp. 90-99). Elsevier.
115. Tuszynski, J. A., Trpišová, B., Sept, D., & Brown, J. A. (1997). Selected physical issues in the structure and function of microtubules. *Journal of structural biology*, *118*(2), 94-106.
116. Uemura, M., Tominaga, Y., Nakagawara, C., Shigematsu, S., Minami, A., & Kawamura, Y. (2006). Responses of the plasma membrane to low temperatures. *Physiologia Plantarum*, *126*(1), 81-89.
117. van Meer, G. (1989). Lipid traffic in animal cells. *Annual review of cell biology*, *5*(1), 247-275.
118. Vasiliev, J. M., & Samoylov, V. I. (2013). Regulatory functions of microtubules. *Biochemistry (Moscow)*, *78*(1), 37-40.
119. Volgyi, K., Juhász, G., Kovács, Z., & Penke, B. (2015). Dysfunction of endoplasmic reticulum (ER) and mitochondria (MT) in Alzheimer's disease: The role of the ER-MT cross-talk. *Current Alzheimer Research*, *12*(7), 655-672.
120. Wang, C., & Youle, R. J. (2009). The role of mitochondria in apoptosis. *Annual review of genetics*, *43*, 95-118.
121. Watson, M. L. (1959). Further observations on the nuclear envelope of the animal cell. *The Journal of Cell Biology*, *6*(2), 147-156.
122. Watson, P. F., & Morris, G. J. (1987, January). Cold shock injury in animal cells. In *Symposia of the Society for Experimental Biology* (Vol. 41, pp. 311-340).

123. Webster, R. E., Henderson, D., Osborn, M., & Weber, K. (1978). Three-dimensional electron microscopical visualization of the cytoskeleton of animal cells: immunoferritin identification of actin-and tubulin-containing structures. *Proceedings of the National Academy of Sciences*, *75*(11), 5511-5515.

124. Witman, G. B. (1990). Introduction to cilia and flagella. In *Ciliary and flagellar membranes* (pp. 1-30). Springer, Boston, MA.

125. Wu, J. (1995). Mechanisms of animal cell damage associated with gas bubbles and cell protection by medium additives. *Journal of Biotechnology*, *43*(2), 81-94.

126. Xiang, C., Wang, Y., Zhang, H., & Han, F. (2017). The role of endoplasmic reticulum stress in neurodegenerative disease. *Apoptosis*, *22*(1), 1-26.

127. Zhou, M., Tu, W. W., & Xu, J. (2015). Mechanisms of microcystin-LR-induced cytoskeletal disruption in animal cells. *Toxicon*, *101*, 92-100.

128. Zoeller, R. A., Morand, O. H., & Raetz, C. R. (1988). A possible role for plasmalogens in protecting animal cells against photosensitized killing. *Journal of Biological Chemistry*, *263*(23), 11590-11596.

# INDEX

## R



## S



## T



## V



## Y



## Z